RIVER BASINS AND COASTAL SYSTEMS PLANNING WITHIN THE U.S. ARMY CORPS OF ENGINEERS

Panel on River Basin and Coastal Systems Planning

Committee to Assess the U.S. Army Corps of Engineers Methods of Analysis and Peer Review for Water Resources Project Planning

Ocean Studies Board

Water Science and Technology Board

Division on Earth and Life Studies

NATIONAL RESEARCH COUNCIL
OF THE NATIONAL ACADEMIES

THE NATIONAL ACADEMIES PRESS
Washington, D.C.
www.nap.edu

THE NATIONAL ACADEMIES PRESS 500 Fifth Street, N.W. Washington, DC 20001

NOTICE: The project that is the subject of this report was approved by the Governing Board of the National Research Council, whose members are drawn from the councils of the National Academy of Sciences, the National Academy of Engineering, and the Institute of Medicine. The members of the panel responsible for the report were chosen for their special competences and with regard for appropriate balance.

This material is based upon work supported by the U.S. Army Corps of Engineers, Humphreys Engineer Center Support Activity under Contract/Grant No. DACW72-01-C-0001 and between the National Academy of Sciences and the U.S. Army Corps of Engineers. Any opinions, findings, conclusions, or recommendations expressed in this publication are those of the author(s) and do not necessarily reflect the views of the organizations or agencies that provided support for the project.

International Standard Book Number 0-309-09220-5 (Book)
International Standard Book Number 0-309-53236-1 (PDF)

Additional copies of this report are available from
The National Academies Press
500 Fifth Street, N.W.
Lockbox 285
Washington, DC 20055
(800) 624-6242 or (202) 334-3313
Internet, http://www.nap.edu

Printed in the United States of America.

THE NATIONAL ACADEMIES

Advisers to the Nation on Science, Engineering, and Medicine

The **National Academy of Sciences** is a private, nonprofit, self-perpetuating society of distinguished scholars engaged in scientific and engineering research, dedicated to the furtherance of science and technology and to their use for the general welfare. Upon the authority of the charter granted to it by the Congress in 1863, the Academy has a mandate that requires it to advise the federal government on scientific and technical matters. Dr. Bruce M. Alberts is president of the National Academy of Sciences.

The **National Academy of Engineering** was established in 1964, under the charter of the National Academy of Sciences, as a parallel organization of outstanding engineers. It is autonomous in its administration and in the selection of its members, sharing with the National Academy of Sciences the responsibility for advising the federal government. The National Academy of Engineering also sponsors engineering programs aimed at meeting national needs, encourages education and research, and recognizes the superior achievements of engineers. Dr. Wm. A. Wulf is president of the National Academy of Engineering.

The **Institute of Medicine** was established in 1970 by the National Academy of Sciences to secure the services of eminent members of appropriate professions in the examination of policy matters pertaining to the health of the public. The Institute acts under the responsibility given to the National Academy of Sciences by its congressional charter to be an adviser to the federal government and, upon its own initiative, to identify issues of medical care, research, and education. Dr. Harvey V. Fineberg is president of the Institute of Medicine.

The **National Research Council** was organized by the National Academy of Sciences in 1916 to associate the broad community of science and technology with the Academy's purposes of furthering knowledge and advising the federal government. Functioning in accordance with general policies determined by the Academy, the Council has become the principal operating agency of both the National Academy of Sciences and the National Academy of Engineering in providing services to the government, the public, and the scientific and engineering communities. The Council is administered jointly by both Academies and the Institute of Medicine. Dr. Bruce M. Alberts and Dr. Wm. A. Wulf are chair and vice chair, respectively, of the National Research Council.

www.national-academies.org

COMMITTEE TO ASSESS THE U.S. ARMY CORPS OF ENGINEERS METHODS OF ANALYSIS AND PEER REVIEW FOR WATER RESOURCES PROJECT PLANNING

PANEL ON RIVER BASINS AND COASTAL SYSTEMS PLANNING[1]

PETER R. WILCOCK, *Chair*, The Johns Hopkins University, Baltimore, Maryland
GAIL M. ASHLEY, Rutgers University, Piscataway, New Jersey
DENISE L. BREITBURG, Smithsonian Environmental Research Center, Edgewater, Maryland
VIRGINIA R. BURKETT, U.S. Geological Survey, Lafayette, Louisiana
JOSEPH J. CORDES, George Washington University, Washington, D.C.
ROBERT G. DEAN, University of Florida, Gainesville
JOHN A. DRACUP, University of California-Berkeley
WILLIAM J. MITSCH, The Ohio State University, Columbus
ROBERT E. RANDALL, Texas A&M University, College Station
A. DAN TARLOCK, Chicago-Kent College of Law, Illinois

National Research Council Staff

JOHN DANDELSKI, Study Director
DAN WALKER, Senior Program Officer
NANCY CAPUTO, Senior Project Assistant

---

[1] The Panel on River Basins and Coastal System Planning was one of four panels, operating under the auspices of a coordinating committee that was convened by the National Academies' Water Science and Technology Board (lead) and Ocean Studies Board to carry out studies mandated in the Water Resources Development Act of 2000. The panel's charge is described in Chapter 1. The panel and staff biographies are provided in Appendix A. The "parent bodies" are listed in Appendix C.

Foreword

In the early 1800s the U.S. Congress first asked the U.S. Army Corps of Engineers to improve navigation on our waterways. From that beginning, the Corps began a program of public works that has reshaped virtually all of the nation's river basins and coastal areas. Today we share in the benefits of those works: a reliable water transportation network, harbors that help link our economy to global markets, previously flood-prone land that is productive for urban and agricultural uses, hydroelectric power, and widely used recreational facilities.

Now, at the beginning of the twenty-first century, the Corps' program is under intense scrutiny. Traditional constituencies press the Corps to complete projects that have been planned for many years and campaign for new projects to serve traditional flood control and navigation purposes. At the same time, environmental and taxpayer groups express concerns about these projects in Congress and in the courts. Some of these groups have exposed technical errors in analyses that have been used to justify projects. For these critics, the Corps' water project development program must be reformed and the budget reduced or redirected.

Some of these same groups are pressing the administration, Congress, and the agency itself toward a new Corps mission, broadly described as environmental restoration. However, the concept of restoration awaits more precise definition, and the science of ecosystem restoration is in its infancy. Nevertheless, it is clear that restoration is a call for water resources management that accommodates and benefits from, rather than controls, annual and multi-year variability in the patterns and timing of river flows and the extremes of flood and drought.

Meanwhile, the Corps is affected by a general trend in all federal agencies toward smaller budgets and staffs. As demands for reform mount, the Corps' current staffing and organization may have to be reconfigured to provide improved and more credible planning reports.

As a result of this national debate over the Corps' programs and the quality of its planning studies, the U.S. Congress in Section 216 of the 2000 Water Resources Development Act, requested that the National Academies conduct a study of procedures for reviewing the Corps' planning studies (Appendix D). In addition, Congress requested a review of the "methods of analysis" used in Corps water resources planning.

In response to this request, the Water Science and Technology Board of the National Academies' National Research Council (NRC), in collaboration with the NRC's Ocean Studies Board, appointed four study panels to assess (1) peer review, (2) planning methods, (3) river basin and coastal systems planning, and (4) resource stewardship and adaptive management, along with a coordinating committee to follow these panels' progress and to write a synthesis report (Appendix C).

Our study panels and coordinating committee held several meetings over the course of the study period beginning in 2001. We spoke with dozens of Corps of Engineers personnel, visited several Corps projects, and heard from different groups with interests in Corps projects. We came away with an appreciation for the dedication of Corps personnel and the complications and challenges they face in trying to be responsive to local project sponsors and the nation's taxpayers.

This is not the first study of the Corps by the National Academies. However, past studies were often focused on specific projects or on particular planning aspects. The reports in this series address the agency's programs in a wider context. Because we appreciate the importance of the U.S. Congress and the sitting administration in directing Corps programs, many of our recommendations are directed to them.

The Corps has a long history of serving the nation and is one of our oldest and most recognized federal agencies, but it is today at an important crossroads. The nation, through the administration and Congress, must help the agency chart its way for the next century.

Leonard Shabman, *Chair*,
Coordinating Committee

Preface

The footprint of the U.S. Army Corps of Engineers (the Corps) on the nation's waterways and coasts is enormous. The Corps has developed and maintains our navigable harbors and waterways, constructed dams large and small, reengineered rivers for flood control, and implemented a diverse range of shore protection measures. The social and economic benefits from flood control, navigation, and erosion protection are enormous, but so too have been the costs, not just for the construction and maintenance of these operations, but for their environmental impacts, cumulative effects, and unintended consequences. It is common, and all too easy, to criticize the Corps for these impacts, although, if examined closely, the criticisms are often made from the perspective of values and objectives that have changed substantially from those in effect when the projects were designed and built. A more useful approach may be to evaluate Corps projects in terms of the objectives specified at the time the projects were built and the authorities and tools available then to the Corps. To be sure, not all Corps projects can be judged a success on these terms. Yet in many cases, the Corps has very effectively achieved the objectives specified for a project, such as providing flood and shoreline protection and reliable shipping channels.

Over the past 30 years, the range of objectives sought for water projects has changed and grown considerably. Much greater value is now placed on environmental and recreational objectives, which serve to increase the complexity of water project planning while also expanding the spatial and temporal scales that must be considered. To meet these demands, the Corps is being asked to undertake integrated water project planning, adopting a watershed or regional approach and including an ecosystem perspective. Integrated water resources planning is widely endorsed by the academic and engineering communities and clearly supported by Corps policy and by public statements of Corps leaders. Although the knowledge and tools necessary to undertake this work are evolving and the record of success is mixed, the Corps has endorsed the challenge and, in some ways, has led the charge.

Effective water project planning in this new environment requires an approach that seeks to balance a diverse range of objectives that cannot be directly or easily compared and to forecast outcomes and impacts of water projects in the midst of the considerable uncertainty inherent in large and complex natural systems. Such efforts are difficult not only because of the complexity of the contemporary multi-objective, multi-stakeholder planning environment, but also because of the complex and conflicting mix of legislation, congressional committee language, administration guidance, and legal precedent that operates as our nation's water policy. The clear policy guidance and consistent funding and authority necessary for integrated planning at the scale of river basins and coastal systems do not presently exist. Integrated water resources planning must also be conducted in competition with strong pressures to build specific projects advocated by local interests and their congressional representatives. Further, even in cases where the need for a comprehensive regional analysis is widely supported, the funding necessary to carry out the analysis may not be available.

Despite these challenges, there is no shortage of examples in which the Corps is successfully engaged in integrated water resources planning and analysis at the scale of river basins and coastal systems. This is not to say that the Corps' efforts in these cases fully satisfy all interested parties; such consensus is unlikely in large-scale, contentious projects with important environmental consequences and a range of stakeholders with conflicting interests. In a regulatory, policy, and political environment that neither fully supports integrated water resources planning, nor is likely to undergo wholesale changes in the near future, the focus of this panel was to evaluate barriers to effective integrated planning at the Corps and to identify changes in its regulations, guidance, and procedures that can help the Corps achieve its new and difficult integrated planning mission within the present political and economic environment.

In developing its report, the panel met three times. At an initial meeting in Washington, D.C., in June 2002, the panel heard from planning experts from Corps Headquarters and set the agenda for its review. At a second meeting in New Orleans, Louisiana, in September 2002, the panel heard presentations from a diverse set of experts from Corps districts, research labs, and the Institute for Water Resources. At a final meeting in Irvine, California, in November 2002, the panel met with members of the other panels and the coordinating committee that, as a group, are conducting a broad evaluation of the Corps' analysis methods and review procedures.

The panel's work was greatly aided by the open, honest, and informed discussions with Corps staff members from all levels: headquarters, division, district, research labs, and the Institute for Water Resources. Although these individuals are also acknowledged elsewhere, it is appropriate to state here that the successful development of this report, and the satisfaction in producing it, can be directly attributed to the highly competent and enthusiastic staff members with whom the panel had the privilege of interacting.

The panel was chaired through August 2003 by Larry Roesner, who provided direction to the panel and liaison with the other panels and whose vision of Corps responsibilities in integrated water planning and environmental stewardship figures prominently in this final report. The panel's work would not have been possible without the support of National Research Council staff. Jeff Jacobs (senior program officer, Water Science and Technology Board, and project director for the three other panels comprising the broader review of Corps planning and review procedures) provided timely and wise advice and assistance. John Dandelski (study director) and Dan Walker (senior program officer) played central roles throughout the panel's deliberations and the production of this report, which simply would not have reached fruition without their good judgment, persistence, and hard work. Julie Pulley (project assistant) ably coordinated meeting logistics and early report drafts, and Nancy Caputo (senior project assistant) was pivotal in producing the final report.

Peter R. Wilcock, *Chair*,
Panel on River Basins and
Coastal Systems Planning

Acknowledgments

This report was greatly enhanced by the participants of the two information gathering meetings held as part of this study. The panel would first like to acknowledge the efforts of those who gave presentations at meetings: Joseph Dixon, U.S. Army Corps of Engineers; J. Craig Fischenich, U.S. Army Engineer Research and Development Center's Environmental Laboratory; Bill Good, Louisiana Department of Natural Resources, Coastal Restoration Division; John D. Kiefer, Kentucky Geological Survey; Kenneth D. Orth, Institute for Water Resources; Russell V. Reed, U.S. Army Corps of Engineers; David V. Schmidt, U.S. Army Engineer District; John Saia, U. S. Army Corps of Engineers; Harry Kitch, U.S. Army Corps of Engineers; Scott Faber, Environmental Defense; Robert E. Turner, Louisiana State University; Robert Brumbaugh, Institute for Water Resources; and Charles B. Chesnutt, U.S. Army Corps of Engineers.

The panel is also grateful to a number of people who provided important discussion and/or material for this report: Arlen Feldman, U.S. Army Corps of Engineers, Davis, California; Brian Moore, U.S. Army Corps of Engineers, Wilmington, North Carolina; Robyn S. Colosimo, U.S. Army Corps of Engineers, Headquarters, Washington, D.C.; and Mark Colosimo, U.S. Department of Agriculture, Natural Resources Conservation Service, Beltsville, Maryland.

This report has been reviewed in draft form by individuals chosen for their diverse perspectives and technical expertise, in accordance with procedures approved by the National Research Council's Report Review Committee. The purpose of this independent review is to provide candid and critical comments that will assist the institution in making its published report as sound as possible and to ensure that the report meets institutional standards for objectivity, evidence, and responsiveness to the study charge. The review comments and draft manuscript remain confidential to protect the integrity of the deliberative process. We wish to thank the following individuals for their participation in their review of this report:

Stephen C. Farber, University of Pittsburgh, Pittsburgh, Pennsylvania
David Ford, David Ford Consulting Engineers, Sacramento, California
James R. Hanchey, State of Louisiana Department of Natural Resources, Baton Rouge, Louisiana
Daniel P. Loucks, Cornell University, Ithaca, New York
Wilbur Mauck (retired), U.S. Geological Survey, Colombia, Missouri
Peter Rogers, Harvard University, Cambridge, Massachusetts
John M. Volkman, Stoel Rives, LLP., Portland, Oregon
Douglas Wooley, Radford University, Radford, Virginia

Although the reviewers listed above have provided many constructive comments and suggestions, they were not asked to endorse the conclusions or recommendations nor did they see the final draft of the report before its release. The review of this report was overseen by Dr. Kenneth Potter of the University of Wisconsin and Mr. Richard Conway (retired) of Union Carbide Corporation, who were appointed by the National Research Council, and were responsible for making certain that an independent examination of this report was carried out in accordance with institutional procedures and that all review comments were carefully considered. Responsibility for the final content of this report rests entirely with the authoring panel and the institution.

Contents

Executive Summary

The U.S. Army Corps of Engineers (the Corps) has played a large and important role in shaping water resources systems in the United States since Congress first tasked it in 1824 to improve navigation on the Ohio and Mississippi Rivers. Since then, rivers have been modified for navigation and flood control, harbors have been dredged for shipping, and coastlines are routinely fortified against erosion and beach loss. Recent decades have seen an overall decline in budgets for civil works project construction, yet the range of objectives for water resources projects has broadened as society places more value on environmental and recreational benefits. Thus, the Corps' portfolio of water resources projects has changed considerably. There is a reduced emphasis on traditional construction projects and an increased focus on maintenance and reoperation of existing projects such as locks, dams, and levees and on environmental restoration projects ranging from local streambank rehabilitation to large and complex projects intended to restore ecosystem function of entire regions.

The expanding range of water projects has increased the complexity of water project evaluation while also increasing the spatial and temporal scale of the necessary analyses. At the same time, the requirement for local cost-sharing and a focus on local client service has pressured the Corps to focus within narrow project boundaries. Water project planning has evolved toward a more collaborative venture, giving voice to many stakeholders representing the diverse objectives that water projects can address. Successful water project planning and evaluation in a multi-objective, multi-stakeholder environment requires an integrated systems approach capable of a balanced evaluation of all relevant issues (e.g., hydrologic, geomorphic, ecologic, social, and economic) over relevant scales of space and time. Such an approach is required to identify unintended consequences, multiple stressors, and cumulative effects and to evaluate trade-offs among competing objectives such that the true costs and benefits of a project may be examined within a context that incorporates the interests of all those with any substantial stake.

An integrated approach to water resources planning at the scale of river basins and coastal systems is widely endorsed by the academic and engineering communities and is clearly supported by the *Unified Federal Policy for a Watershed Approach to Federal Land and Resource Management* (65 Fed. Reg. 62566, October 18, 2000). The Corps' mission, expertise, and experience give it immense potential to alter the structure and functioning of the nation's waterways and coasts. The Corps has embraced its responsibility to plan, develop, and operate water resources projects in a way that considers both economic performance and opportunities for environmental restoration, while minimizing unwanted or negative impacts to other areas within a watershed, adjacent watersheds, and the coastal system. To accomplish this, the Corps has made significant policy changes and has adopted an integrated watershed or regional perspective and environmental stewardship as primary institutional objectives. As might be expected in a large and complex organization answering to a range of public and private demands, implementation of these new policies and objectives is neither consistent nor complete.

At the request of Congress, the National Academies' National Research Council (NRC) appointed four study panels and a coordinating committee to recommend improvements in the Corps' water resource project planning and review process (Appendix D). These panels considered different dimensions of Corps planning (peer review; analytical and planning methods; river basins and coastal systems; adaptive management) (Appendix C). The chairs of the four study panels were all members of the coordinating committee, which enabled it to follow discussions within and among study panels. Each panel operated independently and in accord with NRC guidelines. The coordinating committee issued its own report, which was also subjected to standard National Research Council procedures. In doing so, it considered the draft reports from the panels (in the case of peer review, the panel's final report was used; see NRC, 2002a), as well as discussions among panels, panel chairs, and coordinating committee members. The studies were organized under the auspices of the NRC's Water Science and Technology Board and Ocean Studies Board, with input from the Board on Environmental Studies and Toxicology and the Transportation Research Board.

The charge to the Panel on River Basins and Coastal Systems Planning (shown in Box ES-1) directed it to examine the challenges in water resource planning that are inherent in the nature of large, complex, natural systems. This report conveys the results of the panel's deliberations and makes public its findings and recommendations.

BOX ES-1
Panel Statement of Task

Review and make recommendations on the Corps' planning, design, operation, and evaluation activities in the context of the nation's river basins and coastal systems. Topics covered will include economic and environmental benefits and costs over a range of time and space scales, multiple purpose formulation and evaluation methods, trade-off analysis, inter-agency cooperation, and the integration of water development plans with other projects in the region.

A SYSTEMS APPROACH

Hydrologic systems, and the economic and ecologic systems that they support, are complex, changeable, and interconnected. Effective water project planning requires an integrated approach that can balance the various benefits and costs of a project, while reducing the possibility that attempts to solve problems in part of the system will cause problems in other parts (NRC, 1999a). The merits of an integrated systems approach are endorsed within the Corps, throughout much of the professional water resources community, and by numerous NRC reports (NRC, 1999a, 1999b, 2000, 2001a, 2002a). The panel evaluated the Corps' experience in performing such analyses in the context of river basins and coastal systems and sought to identify changes in Corps regulations, guidance, and procedures that can help it achieve this expanded planning mission.

Toward Improved Integrated Water Planning

An ideal environment for fully integrated water project planning that addresses social, economic, and environmental objectives at all relevant spatial and temporal scales would require a substantial amount of advance investigation and planning at the scale of river basins and coastal systems.

Given adequate time, funding, and authority, such a process could presumably define and evaluate the many trade-offs among competing objectives at large scales of space and time, leading to clear planning guidance at the project scale. Water policy in the United States, however, is following a trajectory away from such central, master plans. Typical Corps water projects are required to have a local sponsor who often advocates a specific project to address a well-defined local need. The sponsor may be unwilling or unable to support the effort required to evaluate the project's role within a broader watershed or coastal system. Although any proposed project must face a complex web of regulatory requirements, conflicting stakeholder interests, and potential legal challenges that serve to communicate some of the broader goals that have been enunciated for our nation's waterways, this process is neither thorough nor efficient. In this context, general policy statements endorsing integrated water systems planning, a watershed approach, and ecosystem restoration may provide little immediate practical assistance for a harried Corps project manager, regardless of his or her inclination to conduct such studies. Current barriers to more effective and consistent implementation of integrated systems planning tend to reflect the limitations of the existing decision-making framework and the presence of conflicting pressures on project planners rather than any unwillingness by the Corps to change.

Constraints Imposed by the Current Project Planning Environment

Congress approves and appropriates funding for the planning and construction of water resource projects. The Water Resources Development Act (WRDA) of 1986 (P.L. 104-303) significantly modified project planning by introducing equal cost-sharing between a local sponsor and congressionally authorized Corps funding. This arrangement has given local sponsors and their congressional delegations a greater role in project selection, design, and most importantly, scope. While this involvement may have made the Corps more responsive to local needs, it has also led to a project-by-project focus or piecemeal approach to water resources development by the Corps. Such an approach can work directly against broader-scale evaluations of water resources and ecosystem needs, with the possibility that undesired impacts and more desirable or equitable projects at a broader scale are missed. Based on recent history, revisions to the planning process introduced by the WRDA 1986 have not eliminated approval of projects with strong local support but widespread criticism on broader economic and environmental grounds.

Efforts to carry out integrated water resources system planning are made more difficult by the complex and conflicting mix of legislation, administrative rulings, and legal precedent that operates as our nation's water policy. River basins and coastal systems typically fall into multiple local and state jurisdictions, and there is no institutional instrument to set policy at a river basin or coastal system level within which large and small water resources projects can be developed and evaluated.

When given the necessary authority and funding (typically following a high-visibility event such as a flood, the listing of endangered species such as the salmon, or the degradation of valued ecosystems such as Chesapeake Bay or the Everglades), the Corps has demonstrated some capacity to carry out multi-stakeholder, multi-objective planning projects that incorporate a diverse range of economic and environmental issues over the necessary spatial and temporal scales. Yet the lack of consistent federal policy guidance and coordinated authority and funding for water resources planning and management has hampered the Corps' ability to consistently plan water resources projects within a broader and integrated systems context.

Water Project Planning Procedures and Analyses

Fully integrated water resource planning and management requires effective guidance to determine appropriate time and space scales of evaluation and to evaluate noncommensurate objectives. Sufficient guidance on integrated planning is not found in current Corps documents, particularly regarding a balanced evaluation of the full suite of social, environmental, and economic objectives inherent in river basin and coastal systems planning and in identifying the appropriate spatial and temporal scales to analyze this diverse range of project objectives. Existing guidance is thorough on traditional benefit-cost analysis, but the heavy reliance on this particular analytical method must be modified in the context of multi-objective, multi-stakeholder integrated studies. Corps planning guidance has not been substantially revised for 20 years and should be updated to provide sufficient and balanced information on how to conduct integrated water systems planning within river basins and coastal systems.

Recommendation: The Corps' planning guidance should be modified to provide Corps planners with contemporary analytical techniques necessary for integrated systems planning on large scales within river basin and coastal systems. Guidelines for identifying all relevant

factors affected by a water project and their spatial and temporal scales, and standards for a balanced evaluation of economic, social, and environmental factors, should be updated and expanded to a level of detail comparable to current standards for traditional benefit-cost analysis of economic objectives of a project (see Recommendations 4-2 and 4-3 in Chapter 4 for more detail).

Project Scoping and Cost Sharing

The local cost-sharing requirement implemented by the WRDA 1986 has affected the Corps in positive and predictable ways and in unanticipated ways as well. Whereas cost-sharing has eliminated a significant number of marginal projects, one major effect has been to limit the scope of project planning studies. The first priority of local sponsors is to ensure adequate monies for addressing the local problem; they may have little interest, authority, or ability to support integrated studies over broader regions (e.g., supporters of a local flood control project may be unable or unwilling to support a regional study of all flood control projects in the same watershed). In contrast, it is clearly a federal interest to evaluate how the project might fit into the broader river basin and coastal system context, in order to identify projects that may prove more beneficial and to identify potential upstream and downstream impacts.

Corps planning studies occur in two phases: an initial reconnaissance study and a subsequent, more detailed feasibility study. Federal support of integrated planning must occur in both phases of the planning process. In the reconnaissance phase (currently 100 percent federally funded), an integrated analysis provides the necessary evaluation of all appropriate benefits and costs at all relevant scales of space and time. Advances in information and decision support technology can support studies of considerable breadth within the reconnaissance study framework.

In the feasibility phase, the portions of the study concerned with a broader evaluation of benefits and costs throughout river basins and coastal systems should be federally funded, while preserving the existing 50 percent cost-sharing for those portions of the project directly concerned with project development, including design, land acquisition, and construction. The requirements for initiating a feasibility study, which currently include a project study plan and a cost-sharing agreement with the local sponsor, should be amended to include the definition of those portions of the feasibility study that constitute integrated river basin or coastal system evaluation and therefore would be federally funded.

Recommendation 5-2: The Corps should ensure that all reconnaissance studies include an integrated evaluation of all project benefits and impacts at any relevant spatial and temporal scale, leading to a definition of the scope and budget of integrated river basin or coastal system analyses required in a feasibility study. Cost-share requirements for feasibility studies should be amended to provide 100 percent federal funding for an integrated evaluation of project benefits and costs at all relevant temporal and spatial scales.

Recommendation 5-3: The scope and budget for integrated planning studies should be determined in the reconnaissance phase and explicitly defined in the project study plan and cost-sharing agreement that define the scope and financial responsibilities of the feasibility study. Approval of a feasibility study should be contingent on a judgment, informed by appropriate internal and external review, that a study plan of the salient social, economic, and environmental factors, at all relevant spatial and temporal scales, has been defined.

Environmental Stewardship

The potential of the Corps to alter the structure and functioning of the nation's ecosystems is significant. As a result, the Corps has a public responsibility to serve as an environmental steward by reducing the environmental risks of its projects and protecting ecosystems against unnecessary and unintended project consequences. The Corps acknowledges that it can make positive contributions to the nation's environment and that it can also irreversibly damage the natural environment. The Corps and other federal agencies have been charged with fostering an "ecosystem approach" that seeks to integrate social and economic goals with the restoration and preservation of natural ecosystems. Simply minimizing harm to the environment is, therefore, no longer sufficient. The Corps should endeavor to improve environmental quality in all of its projects (not just its restoration projects). Potential impacts should take into account not only the spatial scales over which cumulative impacts could occur, but also the time frame over which they might occur.

Recommendation 3-1: The Corps should ensure that all project plans include an assessment of how the project fulfills the Corps' commitment to environmental stewardship. The cumulative effects of each project, together with other past and future human activities in

the same river basin or coastal system, should be consistently evaluated for all projects.

Project Evaluations and Adaptive Management

Uncertainty is an inherent part of the management of all natural systems and its consequences are particularly obvious when ecological outcomes are added to the list of project objectives. The inherent uncertainty in complex and heterogeneous natural systems increases significantly with increasing areal extent. To accomplish resource planning and management in the face of such uncertainty, the concept of adaptive management has gained increased acceptance by the Corps and other land resources agencies. As pointed out in the recent NRC report *Adaptive Management for Water Resources Project Planning* (NRC, 2004a), successful implementation of adaptive management will require resources to support its various components, including monitoring and related science programs, support staff, and stakeholder participation. That report concluded that Congress should support adaptive management within the Corps by providing a consistent level of funding and personnel resources that will allow for the execution of a long-term program. In particular, Congress should broadly authorize and appropriate resources to promote ecosystem and economic monitoring and ex post evaluations of Corps projects.

Determining project success and identifying unsuccessful project components are essential parts of the learning process that can improve the planning and design of future projects. This is particularly true of projects involving large, complex river basins and coastal systems. Although evaluations typically occur at the end of a project, complex projects with higher levels of uncertainty can benefit from ongoing assessments throughout the life of the project. Results of these assessments could be used to adaptively manage the project and inform changes in the project's design. Consequently, the panel strongly endorses the findings reported in *Adaptive Management for Water Resources Project Planning* (NRC, 2004a) and *Analytical Methods and Approaches for Water Resources Project Planning* (NRC, 2004b). The federal government and local sponsors have to accept the necessity of this basic approach to land and water resource projects; evaluating the results of Corps projects is a central component of successful project development, together with improved learning about how the effects of a project cascade through river and coastal systems in both space and time.

Effective project evaluations require flexibility in the scope of the evaluation (not all projects require an extensive evaluation) and should be explicitly defined, and cost-sharing agreed to, in the feasibility phase. Because the complexity and potential consequences will vary from project to project, the current cost limits on post-project evaluations should be replaced with a flexible system in which the scope, tasks, standards, and costs of post project evaluations are determined on a case-by-case basis, as part of a feasibility study. The decision to move ahead with a project should be contingent on the judgment, informed by appropriate internal and external review, that the post-project evaluation plan is sufficient to document the achievement of project objectives, as well as identifying unintended consequences and undesired cumulative effects associated with the project.

Recommendation 5-5: The Corps should ensure that post-project evaluations are a component of all projects and that these evaluations are cost-shared with the local sponsor. The scope, timing, spatial and temporal scale, and funding for these evaluations should be determined during the feasibility study.

OTHER INSTITUTIONAL ISSUES

Watershed planning requires cooperation with multiple agencies at federal, state, and local levels of government, as well as organized groups of stakeholders. Pre- and post-project monitoring and characterization require efforts similar to those of other federal, state, and local entities charged with resource management and environmental protection. These two points suggest that there is a potential for cost savings if data gathering efforts are collaboratively planned and executed and if the results are shared broadly. The alternative is either a continued lack of adequate information or a significant overlap of effort. The panel received conflicting reports of the Corps' ability to engage in formal collaboration with other federal agencies. Some speakers reported that interagency collaboration was feasible and even routine. Others reported that the Corps' funding authorities limit its partnering with other federal agencies, particularly in cases where the Corps is not the lead agency. Review and standardization of procedures are needed for transferring funds and personnel and for developing collaborative memoranda of understanding (MOUs) or memoranda of agreement (MOAs) with other agencies and nongovernmental organiza-tions, particularly when those agencies have controlling authority.

Recommendation 5-4: Congress and the Executive Branch should take the steps necessary, including a standardized procedure for cooperative agreements and MOUs or MOAs, to ensure that the Corps is able to work effectively in collaborative planning and management.

The current state of knowledge is so vast in the water resources area that it can be difficult for one agency to harbor the full range of scientific, engineering, and socioeconomic skills that might be required on a particular project. At the same time, the Corps, like other federal agencies, is currently losing significant parts of its institutional and core competency to retirement. As the water resources planning environment becomes more complex and diverse, the Corps must seek to find a useful balance of in-house and outside expertise. The Corps will have to recruit and train its employees in fields in which they currently lack sufficient expertise. The Corps has reemphasized water resources planning as a key area of expertise and has instituted training initiatives that are vitally important. These initiatives should be consistently supported to provide effective in-house capabilities and incentives for the retention of strong employees. At the same time, the Corps should take advantage of outside scientists and engineers who can bring specialized knowledge or a detailed understanding of the project area.

Recommendation 5-6: The Corps should undertake an effort to review current staffing practices and, if necessary, expand these practices to maintain a well-trained in-house staff and to employ the services of outside scientists and engineers who can bring specialized knowledge or a detailed understanding of the project area.

LOOKING AHEAD

An ideal water planning environment—or even a reasonably good one—will require the support and cooperation of Congress, the executive branch, and the American people and will take time to develop. Although general policy guidance mandating watershed, regional, and ecosystem analysis is clear and publicly supported by current Corps leadership, political support for true watershed or coastal systems planning has been neither consistent nor unanimous. Changes in planning guidance and institutional procedures of the Corps can allow it to effectively and consistently perform integrated water resources planning and environmental stewardship in a river basin and coastal systems context. Effective changes need not require wholesale—and politically controversial—changes in the Corps' organization or in its relations with local clients and federal sponsors.

1

Introduction

When we try to pick out anything by itself, we find it hitched to everything else in the Universe.
—John Muir, *My First Summer in the Sierra*, 1911

WATER RESOURCES SYSTEMS PLANNING AND MANAGEMENT IN THE U.S. ARMY CORPS OF ENGINEERS

Much of the nation's water resource infrastructure was put in place by the U.S. Army Corps of Engineers (the Corps). The Corps' historic mission has been to plan and construct projects with well-defined economic benefits, such as levees, dams, dredged channels and ports, and beach protection. The environment in which the Corps now operates is considerably different and its mandate has changed over the past three decades. The nation has significantly increased its expectation for diverse benefits (at a minimal cost) from public works projects of all types. Water resource projects are discussed and planned in a diverse setting in which authority is dispersed among federal, state, and local governments. Stakeholder groups representing a wide range of interests are being given increasing influence in project planning and approval efforts. Environmental restoration and environmental mitigation now constitute a large and increasing portion of the Corps' work, making the suite of objectives that must be considered larger and more diverse.

These changes are most vividly evident in the context of earlier Corps projects, such as the Missouri River dam and levee system and the Kissimmee River restoration, which are now being modified or removed to satisfy a broader suite of objectives than those for which they were originally designed. Increases in the number and complexity of project objectives, and the diverse regulatory and jurisdictional settings in which Corps projects occur, have increased the complexity of project planning and management.

Today there is less public and congressional support for large water projects and increased support for smaller projects and aquatic system restoration. There can be a tension between the construction of smaller water projects and the broader goals of ecosystem restoration and integrated basin planning, which can require evaluation of a range of factors over large scales of space and time that exceed those historically associated with the development of a small water project. The complexity of modern water resources problems and increasing recognition that the impacts of a water project may extend well beyond its immediate boundaries emphasize that effective water resources planning and management require an integrated approach, an approach able to account for a wide range of objectives and consider a wide range of temporal and spatial scales.

Integrated and system-wide water resources project planning in the United States has always suffered from a mismatch between the nature of physically (and naturally) defined watersheds (where the boundary follows a topographic divide) and the dimensions of relevant political jurisdictions. A basic problem confronting integrated water resources planning and management in river basins and coastal systems has been that these are based on scientific, rather than legal concepts. As a result, it can be difficult to achieve integrated objectives through the existing legal and policy framework in which the Corps operates. Thus, efforts to accomplish large-scale, integrated planning at the river basin and coastal system scale must be superimposed on an existing statutory structure and history that often frustrates such efforts. Further, watershed protection efforts must overcome fragmented, incomplete, and shared regulatory frameworks that exist throughout the three levels of government as well as in the existing allocation of water and land entitlements.

Efforts to transcend political and institutional boundaries in developing a rational integrated approach to water resources planning have a long history with mixed success. The concept of integrated river basin development, which arose in the early twentieth century, was intended to tame, utilize, and protect against the dangers of natural hydrologic variability. Water management organizations, defined by hydrologic boundaries, would employ information from fields as diverse as hydrology and economics to rationalize project choices. In the River and Harbor Act of 1927 (44 Stat. 1010), Congress authorized the Corps to undertake comprehensive surveys to formulate general plans for the most effective improvement of navigable rivers and their tributaries. These surveys came to be called "308 reports," after House Document Number 308, and represent the nation's first comprehensive river basin plans (NRC, 1999b). Ultimately, the ideal of developing a river basin authority similar to the Tennessee Valley Authority (TVA) on all the nation's major rivers was not accepted by Congress and the states. Nonetheless, as more complex projects were contemplated, design requirements eventually stimulated the idea of rational, integrated planning and culminated in the robust period of dam building that followed World War II.

Modern water resources planning developed rapidly in the 1950s and 1960s in association with the construction of new dams and other water resources projects. Engineers dominated the emerging discipline of water resources planning as it became an important academic subject (Maass et al., 1962). In 1955, the Rockefeller Foundation funded the Harvard Water Program, a joint water resource system design seminar for graduate students and government personnel, and the Corps provided support for the program from 1961 to 1965 (Hufschmidt, 1966). More importantly, the idea of an optimum mix of technically supportable projects within one river basin formed an important component of the Corps' culture.

In 1965 Congress passed the Water Resources Planning Act (WRPA) (79 Stat. 245), the high-water mark of federal commitment to integrated, rational water resources planning. This act attempted to centralize federal water resources planning and policy formulation and created a three-part planning approach to national water resources management, to be administered by the U.S. Water Resources Council (WRC) and regional river basin commissions. Water projects were to follow evaluation practices set forth by the WRC. During its tenure, the WRC completed two national water assessments (termed "level A" planning). The WRC also formed river basin commissions and made efforts to develop (but never issued) guidelines for large-scale watershed planning. The principal legacy of the WRC is the creation of the *Principles and Standards for Planning Water and Related Land Resources* (otherwise known as the *Principles and Standards* or *P&S*) for planning specific "level C" water and related land resources projects (38 Fed. Reg. 24,778-24869, September 10, 1973).

In the *P&S*, the WRC endorsed the creation of a new type of federally chartered river basin corporation that would have planning, construction, and regulatory functions, further supporting the principle that the river basin is the correct management unit for integrated, rational water planning. The WRC also called for careful review of all federal water projects, especially interbasin transfers, and for the creation of "an independent review board . . . to keep a check on the project evaluation biases of the Federal construction agencies" (38 Fed. Reg. 24,778-24869, September 10, 1973).

In September 1982, the WRC voted to repeal the existing principles, standards, and procedures (18 CFR, Parts 711, 713, 714, and 716). These were replaced by a new guidance document the *Economic and Environmental Principles and Guidelines for Water and Related Land Resources Implementation Studies*, commonly called the *Principles and Guidelines* or *P&G*, which was approved by President Reagan in 1983 (WRC, 1983). After replacing the *P&S* with the *P&G*, the WRC was disbanded by executive order in 1982 and some of its functions were transferred to the Federal Emergency Management Agency. The *P&G* text was also moved from the "Rules" section of the *Federal Register* to the "Notices" section, thus downgrading this text from legally required rules for planning, as was the case with the *P&S*, to simply guidelines for planning.

Today, integrated river basin planning exists in a more diffuse and ad hoc form, often driven by concerns with environmental restoration. Construction projects continue to have a role in this vision, but they are more likely—as Corps budget figures show—to involve the restoration of degraded aquatic ecosystems. In addition, any project proposed tends to be more contentious than in the past because many more stakeholders claim a role in deciding the future of a watershed and its landscape. The early idea of rational, comprehensive river basin planning and development was primarily a dialogue between expert hydrologists and engineers. Today, this model has been replaced by a more grassroots, stakeholder participation model of watershed governance. As river basin planning moves ahead and the number of stakeholders in Corps projects increases, no one level of government has been able to effectively mediate these multiplying competing interests.

In this diverse, multi-stakeholder environment, the Corps' efforts to promote integrated water resources planning increasingly encounter legal and jurisdictional factors over which it has little control. Land and water laws fragment the watershed landscape into discrete parts. Water rights laws and claims by private users further complicate matters and can lead to the removal of water from the watershed for consumptive use (drinking water and agricultural irrigation). Land law and land-use rights rest initially with private or public landowners, but any land uses that impair water quality or biodiversity are increasingly regulated by government. Water quality laws and regulations remain concentrated at the federal government level with allowances for stricter state or local standards.

Any direct federal role in land-use planning beyond grants and the management of public lands has been fiercely resisted. Not only does the Corps lack control over land uses that impact its watershed activities, but it also does not have control over water rights, which reside with the states. Congress has the power to preempt state law, but this power is seldom invoked. The Corps must work within state water laws except in the relatively rare cases where state law conflicts with its core navigation enhancement and flood control missions.

The Corps' role in water resources planning and management in river basins and coastal systems has become increasingly varied and complex, especially when compared to the era in which much of the nation's water resource infrastructure was built. In the future, the Corps may be required to modify existing facilities in response to actions under the Endangered Species Act and may find that these conflict with its obligations to provide navigation, as was the case in the summer of 2003 on the Missouri River. The Corps may be called on to assist or coordinate efforts at local and state levels to promote watershed restoration. In some cases, it may have to provide technical advice in a supporting role, by virtue of its control of a river's plumbing (e.g., the Missouri River); in other cases, it may act as the lead agency in proposing new basin-wide options. In all instances, incorporating an integrated approach to water resources planning and management would better address diverse and competing objectives, promote the federal interest in environmental stewardship, and balance competing interests.

ONGOING AND RECENT EFFORTS BY THE NATIONAL ACADEMIES TO ADVISE THE NATION REGARDING WATER RESOURCES PLANNING

Controversies and challenges surrounding the Corps' analytical techniques continue. Debate by the U.S. Congress (and many other groups and individuals) continues regarding the appropriate vision of the nation's river and coastal systems and the appropriate role of the Corps. As part of this debate, the U.S. Congress requested the National Academies to study the Corps' review procedures and its methods of analysis (in Section 216 of the Water Resources Development Act [WRDA] of 2000 [P.L. 106-541, 106th Congress]; see Appendix C for text of Section 216).

In response to this congressional mandate, the National Academies' National Research Council (NRC) appointed four study panels and a coordinating committee to perform these reviews. These panels considered different dimensions of Corps planning (peer review; analytical and planning methods; river basins and coastal systems; adaptive management). The chairs of the four study panels were all members of the coordinating committee, which helped the committee to follow discussions within and among the study panels. Each panel operated independently and in accord with NRC guidelines. The coordinating committee issued its own report, which was also subject to standard NRC procedures. In doing so, it considered the draft reports from the panels (in the case of peer review, the panel's final report was used; see NRC, 2002a), as well as discussions among panels, panel chairs, and coordinating committee members. The studies were organized under the auspices of the NRC's Water Science and Technology Board and Ocean Studies Board, with input from the Board on Environmental Studies and Toxicology and the Transportation Research Board.

SCOPE OF THIS STUDY AND ORGANIZATION OF THE REPORT

The Panel on River Basin and Coastal Systems Planning Panel is one of four study panels charged with reviewing the Corps' analytical approaches and methods for implementing water resources projects. As defined in the Water Resources and Development Act 2000, this includes projects for "navigation, flood control, hurricane and storm damage reduction, emergency stream bank and shore protection, ecosystem restoration and protection, or any other water resources project carried out by the Corps." The specific statement of task for this panel, as stated earlier in Box ES-1, follows:

> Review and make recommendations on the Corps' planning, design, operation, and evaluation activities in the context of the nation's river basins and coastal systems. Topics covered will include economic and environmental benefits and costs over a range of time and space scales, multiple purpose formulation and evaluation methods, trade-off analysis, inter-agency cooperation, and the

integration of water development plans with other projects in the region.

Stated broadly, this panel's objective is to evaluate the Corps' efforts in water project planning within the context of river basin and coastal systems and its use of an integrated systems approach to planning within these systems.

Although there are various definitions of a systems approach in the context of water resources, the essential function of a systems approach is to provide an organized framework that supports a balanced evaluation of all relevant issues (e.g., hydrologic, geomorphic, ecologic, social, economic) at appropriate scales of space and time. Within a systems framework, multiple stressors can be identified and quantified, multiple goals can be investigated, trade-offs among competing objectives can be evaluated, potential unintended consequences can be identified, and the true costs and benefits of a project can be examined in a context that incorporates the interests of all those with any substantial stake.

Systems analyses in one form or another have been a part of Corps planning procedures for most of the previous century, although the elements included in the analysis and the methods used to support the analysis have changed in response to shifting public values and advancing science and technology. The merits of a systems approach are broadly endorsed, within the Corps, throughout the water resources community, and in several NRC reports (NRC, 1999a, 1999b, 2000, 2001a). A systems framework supports a balanced consideration of all relevant aspects of water resources problems at all relevant time and space scales. Without this, it is not possible to confidently choose the most favorable solution to a water problem; to identify the unintended consequences of a water project; or to reliably project the long-term consequences of following a specific course of action.

Systems analyses are inherently multi-scale. The appropriate spatial and temporal scale is not the same for all projects. Further, a range of scales must often be considered in planning individual projects, depending on the questions asked. For example, the physical design of an urban stream restoration may be based largely on local factors, although the motivation for the project may be derived in large part from regional considerations regarding the yield of sediment and nutrients from the watershed. Similarly, the specific design of a flood mitigation project may depend primarily on local conditions of flow and runoff, although the impact of the project may have to be evaluated in terms of the cumulative effect of all projects in the river basin.

Systems analyses are also inherently multi-disciplinary. Social and economic values play a clear role in determining the benefits and costs of water projects and in evaluating alternatives. The merits of a water supply project cannot be evaluated adequately without consideration of alternatives to increased water storage, such as conservation measures or developing an interconnected, coordinated, and often cross-jurisdictional regional supply system. The merits of a navigation project cannot be evaluated without forecasting future demand for the waterway and considering the relative merits of alternative forms of transportation. The social and economic benefits and costs of water projects have long been a standard part of the Corps' planning procedures, and specific guidelines for such analyses are part of the Corps' planning portfolio. The existence of these guidelines does not ensure that these evaluations are always judged a success by all interested parties. Political, financial, and provincial factors can exert pressures on decisions made regarding the appropriate scale of analysis, as well as on the costs and benefits assigned.

Over the past three decades the value of the natural environment has played an increasing role in the definition and evaluation of Corps projects, such that the range of disciplines included in water planning has been adjusted to place environmental sciences on an equal and sometimes dominant level. Environmental restoration has become a primary objective—if not the sole objective—of many Corps projects. Although general guidance can be found regarding the need to evaluate the environmental impacts or benefits of a project, guidelines for environmental evaluation lack the specificity of those for determining the economic benefits and costs of a project. Useful information regarding the environmental aspects of projects—particularly the ecological implications of project alternatives—is more difficult to define and more diffuse relative to the economic benefits of a project. Effective trade-offs between environmental and economic benefits and costs remain a difficult, contentious, and unsolved problem. Environmental restoration and environmental mitigation of existing and new projects will likely continue to grow in importance and will magnify the need for effective, integrated planning within river basins and coastal systems.

Human activities that alter the function of various systems are not limited to activities planned and implemented by the Corps. State and local projects and land-use practices can have a significant impact on watersheds and coastal systems and on how these systems respond to Corps-implemented projects. Although it would be difficult to account for all of these activities during project planning, the Corps already has strong ties to many local projects and land-use practice through its regulatory functions.

One of these regulatory functions is to issue permits to state and local land-use projects. In 2003, the Corps issued more than 85,000 permits (U.S. Army Corps of Engineers, 2003a). A fuller accounting of the potential impacts of these permitted activities in any systems analysis may offer greater insight into the cumulative effects of such activities within a given watershed or coastal system and, hence, lead to more effective project design and implementation.

A review of systems analysis in Corps water project planning must also consider the context of broader water policy in the United States. A systems-based evaluation of water projects does not occur spontaneously, but arises in response to some kind of social mandate. The scope of the systems analysis—the range of factors considered, the spatial and temporal scales—depends on the value placed on different water resources objectives. The choices made among the inevitably conflicting objectives will require prioritized guidance. U.S. water policy exists in fragments of nonbinding guidance, environmental law, and standard practice. The absence of a coherent policy, or of a body to develop and interpret that policy, hinders the attempts of the Corps and other water management agencies to plan, design, operate, and evaluate water projects within a genuine and effective systems context.

Chapter 2 begins by discussing the nature of river basin and coastal systems, focusing on the interconnected nature of these systems and the manner in which Corps water projects alter the system's behavior. It summarizes the Corps' influence over water resources systems during the past two centuries and reviews the type, scale, and impact of Corps activities. The Corps' dual role as a regulatory agency and as a civil works construction agency is discussed, and several case studies help to illustrate the scope and impacts of the Corps' activities on inland and coastal waters.

Chapter 3 focuses on the environmental aspects of Corps projects, because these constitute an increasing portion of Corps activity, compel the need for an integrated approach to water project planning and management, and often dictate the spatial and temporal scales requiring investigation. A variety of issues, including knowledge gaps, environmental uncertainty, and the difficulty of balancing environmental and economic objectives, make environmental restoration and stewardship the most difficult challenge facing the Corps in implementing rational, integrated water project planning, as well as its regulatory programs.

The Corps has considerable experience in performing water resources systems planning and has made important contributions to developing integrated water resources planning and management methods. Chapter 4 describes the Corps' current approach to integrating projects within a river basin and coastal system framework, including its governing authorities, methods, and practices. Examples are presented in which an integrated approach to planning and management has been used and barriers to a more consistent application of this approach are identified.

Chapter 5 considers a range of issues (knowledge, jurisdictional, institutional, regulatory, funding, guidance, and policy) that present barriers to a more consistent implementation of integrated water resources management in the river basin and coastal system context. Actions or directions to address these issues are proposed.

2

River Basins and Coastal Systems: The Primary Domains of Integrated Water Resources Project Planning

Chapter Highlights

Planning, construction, and operation of water projects within our nation's river, coastal, and estuarine systems are influenced by a wide range of hydrologic, geologic, geochemical, social, ecologic, economic, and political factors. Simple solutions emerge only when project objectives are defined in terms of a very limited set of these factors. However, water projects usually have more than one purpose, with impacts that cascade among factors and interactions that are often difficult to predict. The primary objectives of this chapter are to delineate the two major types of water resources systems (river basins and coastal systems) and to illustrate the interactions among physical, chemical, biological, and social components of water resources systems over a wide range of temporal and spatial scales. The complex, multi-scale interactions among the aforementioned factors provide the essential rationale for integrated water resources project planning.

LAYING THE GROUNDWORK

A basic tenet of water resources management is that hydrologic systems are interconnected, requiring that effective water project planning take an integrated approach; this approach reduces the possibility that attempts to solve problems in one realm, or subsystem, will cause problems in another (NRC, 1999a). Incorporating a broader view of natural systems in water resources project analysis will increase the nation's economic productivity and environmental well-being in a sustainable manner by minimizing the potential that project benefits in one location are offset by adverse impacts (costs) on other components of the system.

A water resource system is defined by this National Research Council (NRC) panel for the purposes of this discussion as "a set of interrelated physical, chemical, and ecologic components of the hydrospheric environment that act upon, or are acted upon, by one another, and by such interaction thereby determine the unity or whole." Although the division is neither exact nor complete, water resources systems can be usefully divided into two major categories: river basins and coastal systems. "Watershed" and "catchment" are terms similar to river basin and are often used to describe smaller drainages nested within a larger river basin. Coastal systems are geographic units of the coastal zone that can be delineated based on their hydrology, geology, biology, or a combination of the three. The complex nature of coastal systems makes their boundaries harder to define, requiring flexibility and local knowledge in defining a workable unit for environmental management. Most coastal systems—especially estuaries—are strongly influenced by upland watersheds; hence coastal system analysis should generally include both the watershed and the coastal environment.

In addition to linkages between river basins and coastal systems, factors relevant to water project evaluation, particularly its economic, social, and ecologic benefits and impacts, may require consideration across river basin and coastal system boundaries. Water resources project planning that integrates the linkages among the physical, environmental, economic, and societal services of hydrologic systems requires a "systems" approach that is both multi-disciplinary and multi-jurisdictional.

RIVER BASINS

River basins define a well-established and widely accepted framework for designing and evaluating water resources projects (Loucks, 2003; NRC, 1999a,b). River basins are drainage or catchment areas that collect precipitation and transport water, sediment, and dissolved constituents downstream within a system of connected river channels (Figure 2-1). Each basin is separated from surrounding basins by a drainage divide that is a

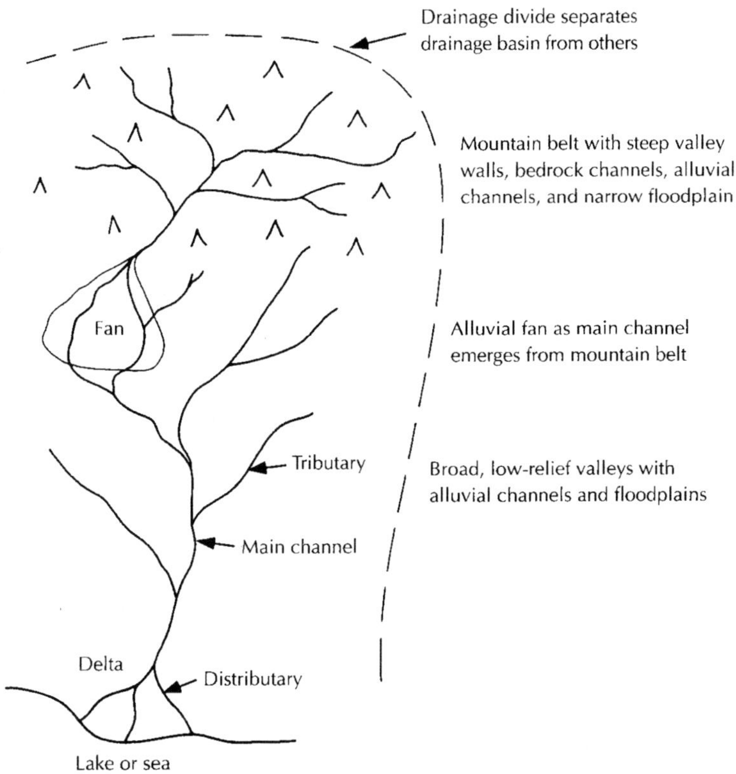

FIGURE 2-1 Typical components of a drainage basin (reprinted with permission from Blackwell Publishing, and John S. Bridge, Binghamton University; Figure 1.1, Bridge, 2003).

physical feature, but not necessarily a political boundary (i.e., political jurisdictions rarely follow such features; consequently, larger river basins typically encompass many federal, state, or local jurisdictions). Nested within this larger physical system are interconnected biological, hydrological, and geochemical subsystems that operate and interact on a variety of temporal and spatial scales, such that changes in one subsystem may trigger changes in others. The river basin concept is a useful framework in which the mass balance of water, sediment, and associated geochemical constituents and their downstream fluxes provide a consistent basis for evaluating system components, their connections, and change.

River basins can be divided into areas of erosion and transport in upstream reaches and transport and deposition in downstream reaches. In the headwaters, the rivers are fast flowing and erosive and channels are typically steep with narrow floodplains. Water projects in this region tend to focus on hydroelectric development and water storage for flood control and for subsequent use in lower parts of the basin. As rivers emerge from the steeper uplands, the fluvial mass balance undergoes a transition from dominantly erosive to dominantly depositional, and river channels grow larger with more extensive floodplains. In this region, water projects focus more on channel works (e.g., locks, dikes, dams, channels) to provide flood protection and reliable navigation. At the distal end of the system (lake or ocean), a delta forms where the river divides into smaller channels (distributaries) and deposits its sediment load. Water projects in this region tend to focus primarily on the removal, through dredging, of sediment deposited in harbors and navigable waterways. At the coast, the freshwater runoff system driven by gravity merges with the saltwater system of the coastal zone driven by waves and tides. The two systems are intricately interconnected; changes in discharge and sediment load upstream in the watershed are ultimately felt downstream in the coastal system.

River basins serve many purposes such as water supply for domestic, industrial, and agricultural purposes; wildlife habitat; transportation and navigation; energy generation; and recreation. Water projects can rarely, if ever, be optimized for all potential objectives that a project might address. Regulation of water velocity and depth in storage reservoirs and channel works for flood control or navigation can alter aquatic and riparian habitat, reduce hydroelectric generating rates, and reduce some recreation opportunities while enhancing others. The spawning and migration of fish, the hydrology of wetlands important to birds and reptiles, and the thriving of plants and animals on different levels of the food chain are all impacted by regulation of normal fluctuations in river water. Evaluation of water projects must account for a large and diverse range of hydrologic, social, economic, and ecologic factors, many of which are difficult to compare directly and all of which depend on a sound understanding of their interaction over a range of spatial and temporal scales.

Watershed Delineation

The term *watershed*, as used in this report, borrows its definition from *New Strategies for America's Watersheds* (NRC, 1999a, p. 39): "a drainage area along with its associated water, soils, vegetation, animals, land use, and human activities." This definition "connotes a relatively small drainage area, while the term river basin is reserved for very large areas. These terms are not scale-specific and should not be limited to particular size classes, however, because each term properly applies to regions ranging in size from less than a small field to almost a third of the North American continent" (NRC, 1999a, p. 34). Each watershed thus has a defining topography with inputs, outputs, and interactions from ecosystem components within that watershed. These open systems may receive water from other watersheds and subsequently transport this water (along with its energy, sediment, nutrients, and contaminants) to downstream watersheds, either naturally or as a result of diversions.

The nation's watersheds have been delineated by hydrologic units that are useful in planning water resources projects. During the 1970s, the U.S. Geological Survey (USGS) and the U.S. Water Resources Council developed a hierarchical hydrologic unit code (HUC) for the United States. The HUC mapping system provides nationwide coverage of surface water drainage with extensive documentation. Attribute tables show hydrologic unit names, numerical codes, and flow direction among cataloging units. The hydrologic units are numerically arranged in a nested fashion, from the smallest watershed (cataloging unit) to the largest (regions; Figure 2-2). During the late 1970's the Natural Resources Conservation Service (NRCS, formerly the Soil Conservation Service) initiated a national program to further subdivide HUCs into smaller watersheds for water resources planning. The 4- to 11-digit HUCs are commonly used by federal and state water resources agencies for water resources inventory and monitoring programs. By the early 1980s, 11-digit HUC mapping was completed for most of the United States. This method of classifying watersheds into progressively finer units provides a basis for addressing the spatial scale of water project evaluations.

Corps Activities in the Nation's Watersheds

Through its Civil Works Program, the U.S Army Corps of Engineers (the Corps) plans, designs, constructs, and operates projects in most of the nation's major watersheds. Project purposes include navigation, flood control, environmental restoration, hydroelectric power, water supply (domestic, industrial, and agricultural), and recreation. The Corps has played an important role in shaping water resources systems in the United States since 1850, when Congress directed the Corps to engage in its first

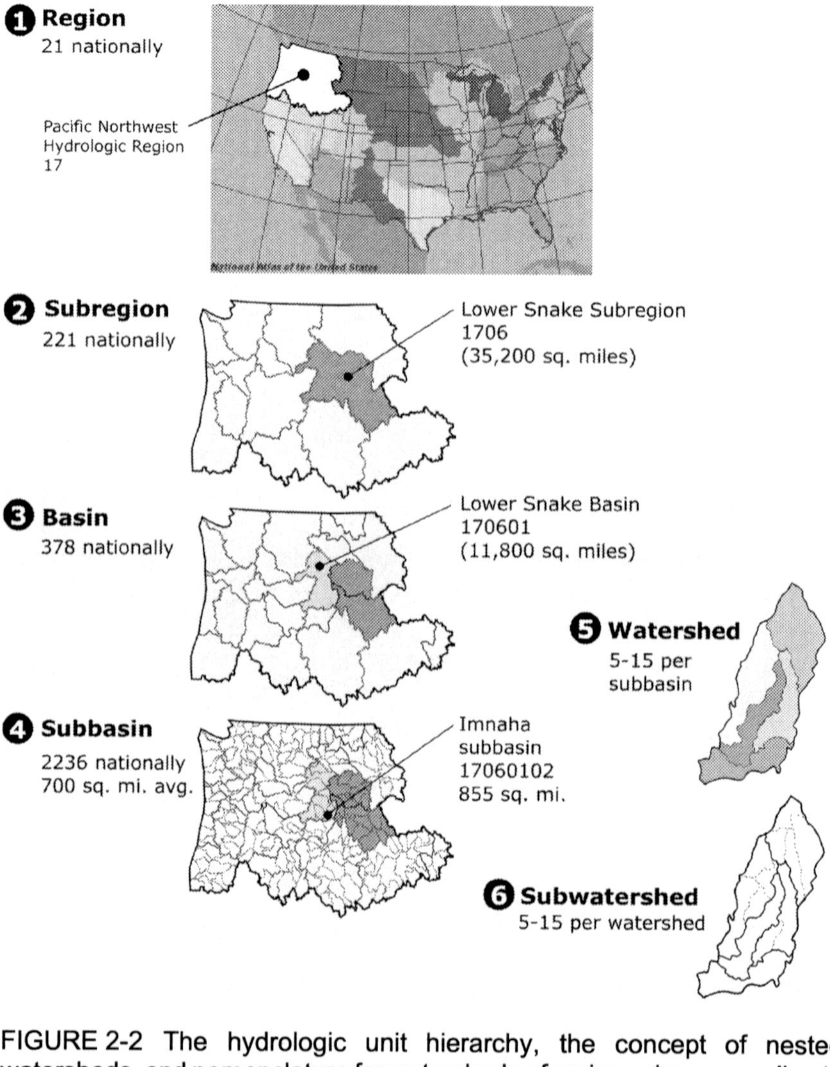

FIGURE 2-2 The hydrologic unit hierarchy, the concept of nested watersheds, and nomenclature for watersheds of various sizes according to the federal hydrologic unit code. Figure courtesy of Bruce McCammon, U.S. Forest Service, Portland, Oregon.

large-scale planning exercise—to determine the most practical plan to control flooding along the Lower Mississippi River (Clarke and McCool, 1996).

The Corps presently operates 384 dams and reservoirs for flood control, water supply, and navigation. The Corps also operates 75 hydroelectric power plants that produce one-fourth of the nation's hydroelectric power or 3 percent of the nation's total electric power supply. The Corps presently maintains more than 12,000 miles (19,200 km) of inland waterways, and it operates 235 locks for navigation. The heart of the nation's inland waterway system—the Mississippi, Ohio, and Illinois Rivers and the Gulf Intracoastal Waterway—accounts for 91.8 percent of the nation's commercial navigation tonnage and is dependent on the Corps for maintenance. The Corps also maintains 300 commercial harbors through which each year pass 2 billion tons of cargo, along with more than 600 smaller harbors (U.S. Army Corps of Engineers, 2003c).

Corps construction projects in inland waterways consist primarily of floodways, cutoffs, revetments, dikes, dams, and channels. Along the courses of major rivers, Corps projects are generally designed to increase the flood carrying capacity of the channel, protect levees, and improve navigation. In smaller watersheds, projects are generally designed to improve drainage, provide water for industrial and municipal use, and increasingly to promote ecosystem rehabilitation.

Understanding the Scope of Water Resource Projects: Examples from the Missouri and Mississippi River Basins

Corps projects in the Missouri and Mississippi River basins illustrate the scale, scope, and type of water resources projects implemented by the Corps under numerous authorizations for flood damage reduction and navigation enhancement in the nation's large river basins. Although comprehensive in some aspects (e.g., hydrologic controls), planning and construction of these projects was, with few exceptions, focused primarily on two functions of these large river systems: conveyance of floodwaters and navigation. The cumulative influence of Corps projects on other functions of riverine systems, such as water quality and wetland maintenance is now receiving increased attention in its analysis of project costs and benefits. The Corps has developed agency-wide environmental objectives that indicate greater emphasis on systems-level analysis of project impacts.

After the 1993 Midwest floods, Congress passed legislation to develop a comprehensive plan for the Upper Mississippi River through the Upper Mississippi River Environmental Management Program and to establish an interstate management council. This legislation required coordination among the Corps, the U.S. Fish and Wildlife Service, and other federal agencies on ecosystem management and restoration as well as flood control and navigation studies. The governors of Illinois, Iowa, Minnesota, Missouri, and Wisconsin created the Upper Mississippi River Basin Association (UMRBA) to fill the void left by the dissolution of a federal-state river basin commission in 1981. In a report defining the strengths and weaknesses of Corps planning (UMRBA not dated), UMRBA recommended that the Corps seek authorizations for large-scale regional programs instead of individual projects.[2]

Case Study: Missouri River. The modern Missouri River basin ecosystem reflects cumulative changes that began with navigation enhancement in the early 1900s and continued with the damming and flow regulation of the mainstem Missouri River in the 1930s. Dam operations have been designed to meet two primary objectives: maintaining navigation and providing flood control. Recreation and irrigation objectives have been subsidiary and generally treated as constraints when designing operating rules to meet the primary objectives. In recent decades, environmental objectives have taken on prominence, particularly through concerns about threatened and endangered bird and fish species. Some Corps activities and their environmental impacts on the Missouri River basin ecosystem are listed below:

- Damming and channelization on most tributary streams, where at least 75 dams have been constructed, have fragmented the river system and partitioned the watershed into smaller units that function somewhat independently of each other, rather than as an integrated whole. Dams have reset water temperatures, trapped sediment (starving downstream areas of a supply of natural sediment), disrupted fish migration, and altered the natural hydrologic variability of many segments of the Missouri River and tributaries (NRC, 2002b).
- By design, the amplitude and frequency of natural peak flows have been sharply reduced by dam construction and managed flows. The replacement of natural, high spring discharges and low summer discharges with steady flows to support barge traffic has altered natural processes. For example, regulated flows (together with sediment trapping in reservoirs) restricted the formation and dynamics of river sandbars, which are crucial to spawning of

[2] Conversely, the UMRBA report cited the success of the Delaware River Basin Commission for its ability to develop strong, legally binding consensus among agencies involved in water resources management.

sturgeon and other fish species and an important habitat for a wide variety of macroinvertebrates and birds. Regulated flows have increased minimum river depths and velocities, increased bed degradation, reduced suspended sediment loads, decreased overbank flooding and associated nutrient supply, and reduced sources of food for wildlife.

- As a result of the nation's effort to manage the Missouri, nearly 3 million acres of natural riverine and floodplain habitat have been altered through land-use changes, inundation, channelization, and levee building. The impacts have reduced biodiversity, affecting species ranging from benthic invertebrates to native fish species to cottonwood trees.

There is an extensive body of scientific research on the Missouri River ecosystem. More than 2,000 studies have been conducted on this ecosystem during the past 30 years, although very few have taken a systems approach that considers the interrelationships between physical, biological, geochemical, and anthropogenic components of the watershed (NRC, 2002b). There is a need for a more integrated planning and management approach that acknowledges linkages between upstream and downstream parts of the basin as well as between various systems or components that make up the Missouri River Basin.

A broader hydrologic issue in Missouri River management is the system-wide effect of levee construction on flood levels. As the proportion of river banks protected by levees increases, thereby increasing the speed with which flood waves propagate downstream, downstream areas could experience increased flooding at the expense of upstream flood protection. Clearly, a system-wide management approach is needed to adequately consider the trade-offs among flood protection, navigation, recreation, and ecosystem services.

The Corps has been the acknowledged controlling authority for Missouri River operations since the completion of construction of the major main stem dams in 1963. In 1979, the Corps codified its operating rules for projects within the basin in the *Missouri River Main Stem Reservoir System Reservoir Regulation Manual* (Master Manual) (U.S. Army Corps of Engineers, 1979). In the 1980s, it became clear that a growing recreation industry, a small shipping industry, and the increased importance attached to environmental impacts of the Missouri water system, required a revision of the operating rules. Since that time, the Corps has been engaged in a long-running effort to revise the Master Manual, which has included extensive modeling of main stem flows, evaluation of benefits and costs among the competing interests dependent on river flow, proposed revisions to the Master Manual, and stakeholder consultation at a range of levels. To date, political differences between upstream and downstream states (particularly regarding the relative importance of navigation, flood control, recreation, and ecosystem restoration) have prevented agreement on revised operating rules and stalled environmental restoration efforts that depend on such revisions.

The unresolved conflicts and absence of a revised master plan came to a head in the summer of 2003 when the Corps found itself subject to two conflicting court orders, one requiring dam releases to maintain depth in the shipping channel and another requiring reduced flows to increase habitat for federally protected bird and fish species. The immediate impasse was resolved in favor of lower flows by a third U.S. court convened to adjudicate the conflict, although a substantial delay in flow reduction was introduced and the long-term prioritization of conflicting objectives remains unresolved.

Case Study: The Mississippi River and Tributaries Project. The Mississippi River and Tributaries (MR&T) Project is one of the most comprehensive Corps endeavors in inland waters. It involves flood control and navigation improvements in all four major river basins of the Lower Mississippi River Valley: St. Francis in east Arkansas; Yazoo in northwest Mississippi; Tensas in northeast Louisiana; and Atchafalaya in south Louisiana (Figure 2-3).

MR&T project work is authorized by the Flood Control Act of 1928 (FCA) and its amendments. The original FCA authorized work that would protect the Lower Mississippi River Valley against Mississippi River floods only, although the tributary streams within the basins caused frequent flood damage and could not be prevented by the main stem Mississippi River protective works. Later amendments to this act authorized work to also alleviate flood problems of the tributaries in all four aforementioned river basins.

In March of 2002, Brigadier General E.J. Arnold reported that the MR&T project was 87 percent complete physically. He also estimated that the nation had invested about $10.8 billion for planning, design, construction, operation, and maintenance of the project and that the accumulated savings in flood damages totaled more than $258 billion (Arnold, 2002).

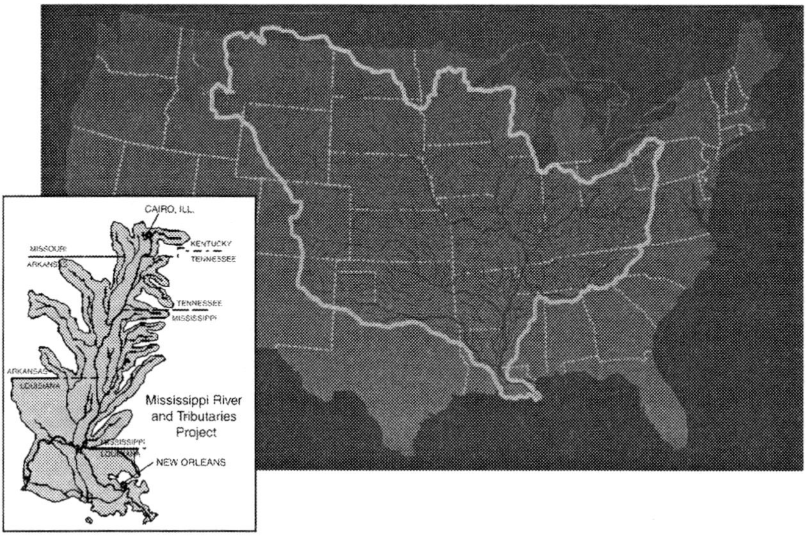

FIGURE 2-3 The Mississippi River Drainage Basin (outlined in light gray). The Mississippi River and Tributaries (MR&T) Project area is the lowest portion of the Mississippi River Drainage Basin, beginning in lower Kentucky and ending at coastal Louisiana (see smaller box, lower left). Figure courtesy of U.S. Army Corps of Engineers, New Orleans District.

The MR&T project is made up of four construction elements: (1) levees for containing flood flows; (2) floodways for the passage of excess flows past critical reaches of the Mississippi; (3) channel improvement and stabilization of the channel in order to provide an efficient navigation alignment, increase the flood carrying capacity of the river, and protect the levee system; and (4) tributary basin improvements, such as dams and reservoirs, pumping plants, and auxiliary channels, for major drainage and flood control .

The Mississippi River levee projects are designed to protect the alluvial valley against the projected flood by confining flow to the leveed channel, except where it enters the natural backwater areas or is diverted purposely into restricted floodways. The main stem levee system, comprised of levees, floodwalls, and various control structures, is 2,203 miles long (Figure 2-4). Some 1,607 miles lie along the Mississippi River itself and 596 miles lie along the south banks of the Arkansas and Red Rivers and in the Atchafalaya basin. The levees were constructed by the federal government

a

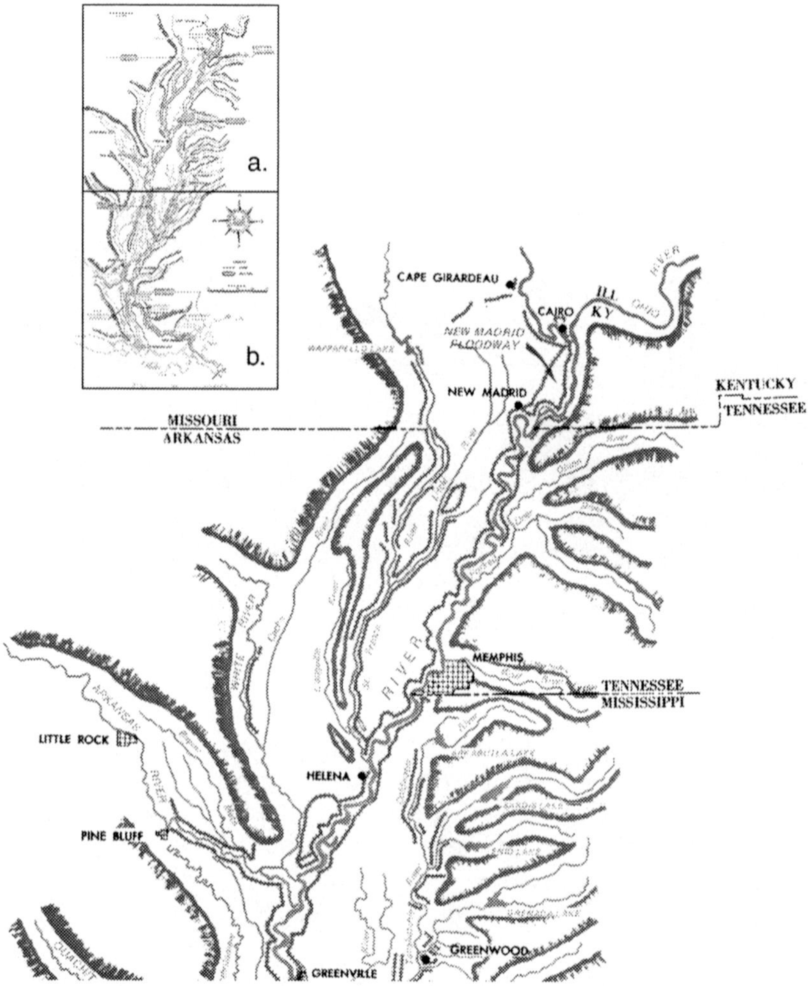

FIGURE 2-4 Major levees and floodways of the MR&T project. Figure courtesy of the U.S. Army Corps of Engineers, Vicksburg, Mississippi.

b

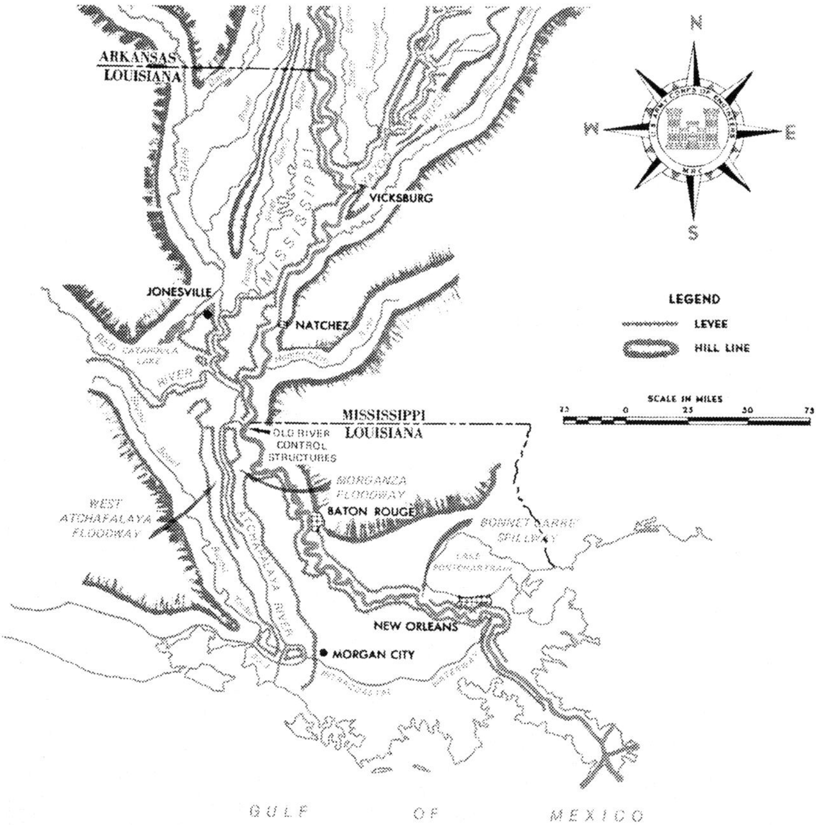
ARKANSAS
LOUISIANA
N
W
E
S
VICKSBURG
MISSISSIPPI
JONESVILLE
NATCHEZ
LEGEND
LEVEE
HILL LINE
SCALE IN MILES
MISSISSIPPI
LOUISIANA
OLD RIVER CONTROL STRUCTURES
MORGANZA FLOODWAY
BATON ROUGE
WEST ATCHAFALAYA FLOODWAY
BONNET CARRE SPILLWAY
ATCHAFALAYA RIVER
NEW ORLEANS
MORGAN CITY
GULF OF MEXICO

and are maintained by local interests, although the Corps periodically inspects them to ensure that they are functioning properly within the MR&T master flood control plan. This plan has led to the construction of 389 drainage structures, 59 pumping stations, 3,731 miles of flood control channels, 3,466 miles of levees, and 44 flood control reservoirs in the Lower Mississippi River Valley (U.S. Army Corps of Engineers, 2002a).

Case Study: Yazoo Backwater Pumping Plant Project. The previous case studies point out the complex and significant challenges facing efforts to systematically evaluate and coordinate project planning and implementation in large and complex river basins. However, large geographic scope is not a requirement for contested water project planning in complex systems. The interaction of two subsystems in a limited area can create significant problems, if not adequately accounted for in project planning. Like the challenges presented in the previous studies, lack of understanding of the impacts of modifying one component of the system can often be attributed to a lack of willingness on the part of local sponsors to investigate the potential benefits of addressing the problem nearer the source or the potential adverse impacts on various subsystems. This case study discusses the Corps plans to build the Yazoo Backwater Pumping Plant in the Yazoo basin, one of the four river basins in the Lower Mississippi River Valley (Figure 2-5). It serves as an example that lack of an integrated, system-level approach to project planning can lead to divisive and unresolved water resources conflicts.

A tributary of the Mississippi, the Yazoo River flows into the mainstem near Vicksburg, Mississippi. Approximately 20 million acres of forested wetland in the Lower Mississippi Valley have been converted to agricultural use (Keeland et al., 1995; McDonald et al., 1979; Tiner, 1984). Today, farming occurs on an estimated 100,000 acres of flood-prone land that have a 50 percent chance of being inundated each year (the two-year floodplain; Shabman and Zepp, 2000). A 1941 authorization for flood control in the Yazoo Basin included provisions for a large pumping station on Steele Bayou that would transfer floodwaters from the backwater area into the Mississippi River (Shabman and Zepp, 2000) in an attempt to prevent flooding of lands in the two-year floodplain.

The purpose of the Yazoo backwater pump and other elements of the 1941 plan was to provide flood protection to a specific level (i.e., to reduce the level of a five-year flood event to 90 feet mean sea level). In 1959, the

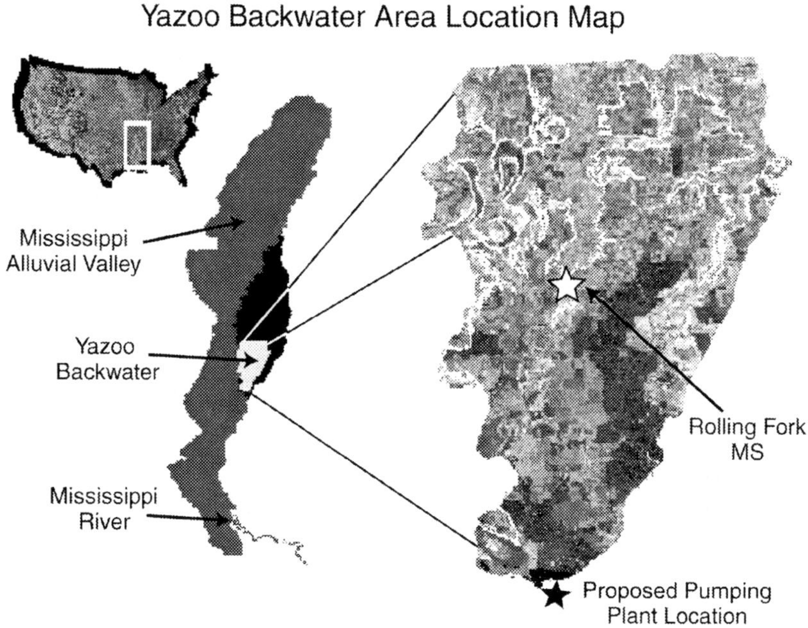

FIGURE 2-5 Yazoo backwater area. Figure courtesy of the U.S. Fish and Wildlife Service, Vicksburg, Mississippi.

Corps concluded that a pumping plant was not necessary to meet this level of flood protection and eliminated that feature from the flood damage reduction plan for the region. The Corps concluded (U.S. Army Corps of Engineers, 1959, p. 20):

> Since the original authorization for Yazoo Backwater Protection, important hydraulic changes have taken place due to improvement of channel efficiency in the Mississippi River and to reservoirs and channel improvements in the Yazoo Basin headwater area. These have resulted in less frequent flooding, and shorter duration of flooding, which makes it feasible to develop a simplification of the authorized plan by eliminating pumping at a large saving in project cost.

The Corps began a reexamination of this Yazoo pumping project in 1991 in response to a request for review and redesign of flood control within the Yazoo Basin, including the proposed Yazoo Backwater Pumping Plant, by the governor of Mississippi and subsequent instruction from the U.S. Office of Management and Budget (OMB) (U.S. Army Corps of Engineers, 2000a). A draft reformulation report and draft environmental impact statement (EIS) on the Yazoo Backwater Pumping Plant Project was released for public comment in September 2000. The final reformulation report and the EIS for the Yazoo pumping plant have not been completed.

The U.S. Fish and Wildlife Service (USFWS) and the U.S. Environmental Protection Agency (USEPA) maintain that more than 200,000 acres of ecologically significant wetlands would be adversely affected by the Yazoo pumping project as proposed in the draft reformulation report (USEPA, 2000; USFWS, 2003). The Corps estimates that only 23,000 acres of wetland would be adversely affected (U.S. Corps of Engineers, 2000b). The presence of pondberry (*Lindera melissifolia*), an endangered plant, in the wetlands that would be drained by the pump is another problem cited by the USFWS (USFWS, 2000a,b).

The USFWS is concerned that the backwater pumping project will drain wetlands already restored by the Wetlands Reserve Program (WRP) and the Conservation Reserve Program (CRP), as well as a portion of Delta National Forest (USFWS, 2000b). Approximately 50,000 acres of marginal agricultural lands in the Yazoo Basin have been restored to wetlands in the drainage area by landowners since 1985, and there is presently a waiting list of landowners who would like to participate in future wetlands restoration under the WRP.

The USFWS and the USEPA have both objected to the project as proposed in the draft reformulation report, and both agencies have recommended that nonstructural flood control measures be implemented rather than construction of the pumping plant. The USFWS has advised the Corps to enter into formal consultation, as prescribed under the Endangered Species Act, on the pondberry issue (USFWS, 2000c). The USEPA's opposition to the project is sufficiently strong that it has threatened to veto the project under its Clean Water Act authority—an action taken only 11 times in the history of the Clean Water Act.

Construction dollars are not generally obligated for Corps projects prior to completion of the EIS required for a project of this scope. In February 2003, however, Congress added $10 million to the Corps' budget for the first phase of construction of the Yazoo backwater pumping project, thereby increasing the stakes in this unresolved conflict.

There is presently no mandate for the joint planning or approval of federal activities with watershed-level-implications in the Yazoo basin. Nor is there an effective mechanism for resolving the conflicting objectives of increased flood control, wetlands restoration, and endangered species protection. Similar conflicts exist in other river basins (i.e., the Apalachicola-Chattahoochee-Flint and the Alabama-Coosa-Tallapoosa River basins; Loucks, 2003). Unless the full complexities of system interactions can be accounted for, conflicts over water resources projects will likely intensify as population growth, economic development, and increased demand for environmental stewardship and recreation alter society's needs for water supply, flood control, and other services within the nation's river basins. Interest in developing a systems-level framework for water resources planning appears to be gaining appeal in some regions, especially as a tool to facilitate planning in interstate river basin systems.

COASTAL SYSTEMS

These case studies from the Missouri and Mississippi River basins demonstrate the complexity of planning and evaluating water projects at a variety of scales within geographically complex river basins. The boundaries of river basins are easily delineated based on topographic divides (as discussed earlier, this allows component watersheds to be managed as units), whereas the complex and dynamic nature of coastal systems makes their delineation more problematic. Unlike river basins that can be defined purely on hydrologic grounds (where surface waters are a unifying component of the various subsystems), coastal systems can be defined on a variety of grounds. For example, the Coastal Zone Management Act of 1972 (P.L. 92-583) defines the term *coastal zone* as "the coastal waters (including the lands therein and thereunder) and the adjacent shorelands (including the waters therein and thereunder), strongly influenced by each other and in proximity to the shorelines of the several coastal states, and includes islands, transitional and intertidal areas, salt marshes, wetlands, and beaches." Thus, attempting to determine which coastal lands and waters may be linked through a single coastal system is less obvious and must be determined on a case-by-case basis.

Despite the obvious challenges associated with attempts to delineate specific coastal systems within a region of the coast using the coastal zone definition just cited, the definition does have utility for management purposes because it clearly recognizes that the various components are linked physically and that interaction with one component can impact the others. Thus, this report uses that definition as a tool for discussing coastal systems in a management context.

Occurring along the narrow border between inland and marine environments, coastal systems are considered highly vulnerable to the effects of human development activities. Although coastal areas comprise only about 5 percent of the world's surface, it is estimated that 50 to 70 percent of the global human population lives in the coastal zone (Intergovernmental Panel on Climate Change [IPCC], 1995). Approximately 53 percent of the total population of the United States lives in 17 percent of the land area considered coastal (Culliton, 1998). Along heavily developed U.S. shorelines, a mosaic of seawalls, jetties, levees, and navigation channels has disrupted natural circulation patterns; altered salinity levels; and changed the rate of freshwater, sediment, and nutrient delivery. The growing demands of society for coastal property and coastal resources have compelled state and local governments to call for navigation improvements, flood protection projects, beach nourishment, and ecosystem restoration projects.

The challenge for water resource project planning and implementation in the coastal zone lies largely in developing the ability to consistently determine and account for the potential system-wide impacts of projects that may involve a single component of a system whose geographic definition may vary depending on the nature of its various components. Understanding this challenge and its complexity forms a significant theme for the following discussion.

Delineating Coastal Systems

A coastal system can be subdivided into the *outer* portion, the portion that is generally under the action of waves generated by winds acting over large expanses of water, and the *inner* portion, the portion that is acted on primarily by tides, locally generated waves, and freshwater inflows. The inner coastal system, which includes bays, estuaries, and adjacent habitats, is generally subjected to less wave energy, and thus, finer sediments are found there. The inner coastal system is often where rivers discharge their sediments in bays (e.g., Chesapeake Bay, Delaware Bay). The Corps' role in managing a coastal system varies based on these subdivisions. The Corps' primary activities in the outer coastal system are to maintain navigation and to control beach erosion; its activities in the inner coastal system involve navigation, flood control, and more recently, ecosystem restoration.

Coastal systems are more difficult to delineate than terrestrial catchments due to the low relief of the adjacent floodplains, the mixing of marine and freshwater inputs in inner coastal waters (which is influenced by winds, tides, basin morphology, and drainage), and the lack of integrated bathymetric, topographic, and circulation data for most coastal watersheds. As a result, coastal systems are rarely identified on maps. Furthermore, the likelihood that a jurisdictional boundary will somehow coincide with the regional extent of a coastal system is even less than it would be for the topographic divide that bounds a watershed. Unlike Corps district boundaries that recognized the value of honoring watershed boundaries, Corps district boundaries in coastal areas (like many of jurisdiction boundaries in these regions) are often straight lines trending at a high angle to the shoreline. Thus, these district boundaries do not reflect the hydrology of these inner coastal systems, where environmental gradients may parallel the shore as frequently as they lie at right angles to it.

The complexity of natural systems that exists in coastal areas is matched by the complexity of government jurisdictions and responsibilities. At least 15 federal agencies collect or use coastal mapping data, not to mention numerous state and local agencies as well as private organizations, and not all of these parties organize and use the data in the same way.[3] Problems have been identified with current hydrologic unit maps in the coastal zone (Ferguson, 2000) and have led to the formation of an interagency Coastal Watershed Working Group of the Federal Geographic Data Committee (FGDC) (U.S. Department of Agriculture, 2003a). A draft proposal including a revised method for delineating boundaries of hydrologic units in inner coastal waters is in review by this FGDC working group. The proposed methodology is integrated with methods for delineating terrestrial watersheds and is based primarily on the delineation of submerged and emergent geomorphic features (e.g., islands, shoals, tidal deltas, inlets, mixing basins).

Understanding the complex nature of coastal systems is not simply an academic exercise. The systematic clarity brought by organizing Corps districts along watershed boundaries is absent along the coastline. Multiple water resource projects within a single watershed are often designed and implemented within a single district. Conversely, water resource projects with a single coastal system stretching from southern New Jersey to North

---

[3] A recently completed report *A Geospatial Framework for the Coastal Zone: National Needs for Coastal Mapping and Charting* (NRC, 2004c) examined the spatial information requirements of the United States and its principal user groups and offered recommendations to improve the nation's mapping needs.

Carolina may involve as many as four separate Corps districts and two Corps divisions.

The first steps toward effective system planning and implementation in coastal systems involve understanding the component subsystems that exist in coastal areas (inner and outer coastal systems) and how these subsystems interact.

The Outer Coastal System

The outer coastal system includes coastal barrier islands, mainland beaches, and the active portion of the continental shelf. The outer portion of a coastal system exchanges sediment with the inner coastal system, and the rates and locations of this exchange may be altered by water projects that modify sediment transport and deposition dynamics. River entrances and tidal inlets, whether natural, modified, or constructed, form a vital part of this sediment exchange system.

Coastal barrier islands are landforms that separate estuaries from the open ocean. They are usually elongated and somewhat parallel to the edge of the land mass and have a non-marsh dry interface (such as sand or rock) with the ocean. Coastal barriers include islands as well as peninsulas connected to the mainland, each providing a barrier to wave action and the mixing of estuarine and ocean waters. Mainland beaches are also considered coastal barriers under the Coastal Barrier Resources Act of 1982 (P.L. 97-348). Longshore currents play an important role in maintaining barrier islands and shorelines by redistributing the available sand from headland areas.

For outer coastal systems, shore stabilization projects that add good-quality sand to the system are generally considered beneficial, whereas features that interfere with natural sand transport (e.g., groins) cause an adverse effect on the adjacent shorelines. Shoreline stabilization activities are usually considered where there is either an erosion trend or a significant erosional event due to a major storm. Erosion trends can have natural or anthropogenic causes or a combination.

Natural causes of beach erosion include currents, storms, winds, waves, tides, relative sea-level rise, and earthquakes. Global sea-level rise (referred to as eustatic sea-level rise) is a major concern of communities located along low-lying coastlines because the present rate of sea-level rise is projected to increase two- to four-fold during the next 100 years (IPCC, 2001). Relative sea-level (RSL) rise is the increase of sea level relative to land elevation and can be due to eustatic rise, land elevation decrease, or a combination of the two. An approximate relationship of shoreline erosion to RSL rise developed by Bruun (1962) states that on a mainland shoreline (i.e., not a barrier island), the shoreline will retreat at a rate of some 50 to 60 times the rate of relative sea-level rise. The average eustatic sea-level rise was approximately 1-2 mm per year during the twentieth century (IPCC, 2001); however, the rate of RSL rise varied substantially among coastal regions due to the withdrawal of subsurface fluids, postglacial isostatic rebound, tectonic activity, and other factors that affect the elevation of the land surface. As an example of the effect of postglacial adjustment, the relative rate of sea-level rise along the east coast of Florida is approximately 0.7 feet per century, approximately 80 percent greater than the eustatic rate (Hicks et al., 1983). The RSL in many locations in Alaska is declining at approximately 1 m (3 feet) per century due to glacial rebound. Another cause of shoreline erosion on portions of the Pacific coast is the presence of submarine canyons that extend quite close to the shore and thus intercept and remove sand from the coastal system and transport it into deep water

Anthropogenic causes of shoreline erosion warrant special consideration here, because, in many cases, engineering solutions can reduce or mitigate adverse effects caused by some Corps projects. Water development projects that have contributed substantially to coastal erosion on the U.S. Atlantic and Gulf of Mexico shorelines are generally those that prevent longshore sediment transport and those that trap sediment (e.g., building jetties, deepening navigation channels, disposing of dredged coastal sand in deep water). The effects of these projects on shoreline integrity can be particularly significant in areas where a strong net longshore sediment transport exists.

On the West Coast, the dominant anthropogenic cause of erosion is the reduction of sand supply from rivers through dam construction and sand mining in riverbeds and directly from beaches. An additional cause of anthropogenically induced beach erosion on the West Coast is the withdrawal of large amounts of ground fluids, water and hydrocarbons, resulting in lowering the ground elevation. A well-documented case is Terminal Island, California, where the withdrawal of hydrocarbons between 1940 and 1958 resulted in lowering the ground elevation by more than 20 feet (Allen and Mayrega, 1970). Present regulations in Southern California require that the withdrawal of hydrocarbons must be accompanied by the injection of an equivalent volume of salt water into the same aquifer.

As the value of coastal property rises, the rationale for improving sediment management practices generally becomes stronger and more appropriate. Much earlier, the Corps' normal procedure was to dispose of sediment dredged from a navigational channel in "the most cost effective manner," which was usually to dispose it offshore in water too deep to benefit the nearshore system. In Florida, more than 50 million cubic yards of sand with a current market value in excess of $500 million has been disposed in this manner (Dean et al., 1988). Some states have enacted legislation to prevent the placement of dredged sand in water that is so deep that little if any benefit to the shoreline results. Nevertheless, some Corps districts continue to place sand in water so deep that either little benefit results or the time required for this sediment to return to the shore is so great that considerable adverse impact to the shoreline occurs in the interim.

Beach and nearshore sediments are a valuable natural resource providing storm protection for coastal infrastructure and recreational opportunities. The natural pathways of these resources are interrupted by deepened navigational channels and their associated jetties, and in some cases, disposal of dredged sediment in water depths too great to benefit the adjacent shorelines. These effects and practices can cause significant erosion to one or both of the adjacent shorelines. Effective management of coastal sand resource requires identification of the sources, pathways, and sinks of sediment within the coastal system.

As development of the nation's shoreline increased during the past 30 years, beach nourishment emerged as the locally preferred option for shoreline stabilization. Earlier methods were predominantly groin construction to trap sand from the littoral drift. This method had the effect of benefiting property where the groins were constructed, resulting in wider beaches but transferring the erosional pressure to downdrift beaches. This approach was implemented four decades ago when much less was known regarding the effects of groin construction and, with the lower development density, downdrift beaches were frequently uninhabited. An additional early approach to shoreline stabilization was through seawall construction; however, although this approach could protect local property against loss, in the presence of a pervasive erosional trend the beach would be lost.

Recognition of the need for an integrated approach to coastal sediment management led the Corps to develop the Regional Sediment Management Program, a large research and demonstration program designed to promote more effective management of this valuable resource.

The Inner Coastal System

The inner coastal system consists of estuaries, interior shorelines, flats, and vegetated wetlands. An estuary is defined as "a partially enclosed coastal body of water which is either permanently or periodically open to the sea and within which there is a measurable variation of salinity due to the mixture of sea water with fresh water derived from local drainage" (Day, 1981, Chapter 1). Many semi-enclosed coastal waterbodies such as bays, inlets, gulfs, and sounds can be classified as estuaries if significant freshwater inflow from river systems occurs. Roughly 80 percent of the U.S. Atlantic and Gulf coasts and 20 percent of the Pacific coast consists of estuaries, with approximately 900 individual estuaries identified (Kennish, 1986). Estuaries are typically fringed with intertidal wetlands, including mangroves and salt-tolerant marsh grasses.

Organisms that inhabit inner coastal environments are uniquely adapted to environmental conditions that occur along the energy, salinity, and moisture gradients that extend from the subtidal region to the uppermost limits of tidal influence (Burkett, 2001). In contrast to terrestrial systems that have physical gradients that can extend over tens to hundreds of kilometers, many coastal gradients are relatively short (Raffaelli et al., 1991). Shoreline configuration and bottom topography of estuaries strongly affect their circulation patterns, and the input of riverine flow has important effects on estuarine water quality. The salinity of estuaries, which controls the zonation of plants and animals, is largely a function of tidal fluctuation and river discharge. Since the condition of an estuary is inherently influenced by upland drainage, water resources projects within a watershed will influence the environmental conditions in an estuary and, in some cases, the condition of the outer coastal system and marine environment.

The natural depths of estuaries are usually too shallow to accommodate commercial and some recreational boating interests, which has led to the dredging of deep draft shipping channels, ports, access canals, and other waterways. Disruption of the inner coastal system is also produced by the construction of shore protection works (bulkheads, riprap) that directly destroy or modify habitats and prevent the natural landward migration of wetlands as sea level rises (Burkett, 2001). Coastal wetlands have also been lost due to dredge-and-fill activities associated with agricultural, municipal, and industrial development. Storage reservoirs can trap sediments and reduce the natural sediment load reaching inner coastal systems. Channelization and leveeing of coastal waterways can also reduce sediment supply to coastal systems by routing the sediment to deep water, as for the Mississippi River. Navigation projects can alter circulation patterns and water quality in inner coastal systems, leading to ecological changes and, in some cases, serving as an obstruction to the migration of coastal habitats as sea level rises.

Estuaries and wetlands are being severely impacted nationwide by human development activities. More than 90 percent of the intertidal wetlands in California have been lost as a result of human development activity (Watzin and Gosselink, 1992). Large areas of intertidal marsh and swamps have been impounded and filled along the Atlantic and Gulf of Mexico shorelines. Human-induced environmental change is pervasive in areas impacted by discharges from sewage and industrial plants, agricultural runoff, and invasive species. An estimated 60 percent of U.S. estuaries are impaired by some form of pollution or habitat degradation (Bricker et al., 1999). Nutrient enrichment associated with agricultural runoff can result in estuarine hypoxia, harmful algal blooms, and fish kills. Eutrophication and habitat destruction are major threats to the healthy functioning of estuaries.

Corps Water Resource Projects and Shoreline Erosion: Examples from U.S. Coastal Areas

Corps projects in coastal areas illustrate the scale, scope, and type of water resources projects implemented by the Corps under numerous authorizations for addressing coastal erosion and navigation enhancement in the nation's coastal zone. These projects also illustrate the complexity of coastal systems and the need for an integrated approach to planning water projects. The off-site impacts of projects include (1) interruption of longshore sediment transport through navigation channel construction or channel deepening, stabilization by jetties, and disposal of sediment in deep water; (2) alteration of salinity, tidal energy, and erosion patterns through changes in the topography and bathymetry associated with channelization and dredging projects; (3) sediment trapping by shore structures such as groins, jetties, and revetments; (4) alteration of sheet flow across wetlands by the construction of levees, roads, and other development projects; (5) introduction of contaminants via dredge spoil, point source discharges, and nonpoint sources; and (6) alteration of geomorphology of tidal inlets through the dredging of waterways and harbors, which can exacerbate erosion in adjacent tidal settings. The variability of coastal systems and how it contributes to these adverse impacts can be illustrated through a number of specific examples. The four case studies that follow show the variability of coastal systems and how Corps navigation projects have impacted ecosystems within these coastal systems.

Case Study: Port of Palm Beach, Lake Worth Entrance. An entrance into the port of Palm Beach, known as the Lake Worth Entrance, was cut in 1918 to provide navigational access to Lake Worth, Florida, and was stabilized by jetties by 1925. The channel was originally only 5 feet deep (Walker and Dunham, 1977), but was deepened gradually to its presently authorized depth of 40 feet. The natural net longshore sediment transport in this area is to the south at an estimated rate in excess of 200,000 cubic yards per year. This transport has been interrupted by the entrance and its jetties, much like a dam, and this has resulted in sediment accumulation on the updrift side and severe erosion on the downdrift side.

For many years, the sand dredged to maintain the authorized depth was placed offshore in water depths too great for waves and currents to return the sand to the nearshore transport system, resulting in the permanent loss of approximately 3 million cubic yards of this high-quality material. Another 3 million cubic yards has been placed in upland sites, and approximately 1 million cubic yards has been impounded by the updrift jetty. This total 7 million cubic yards of sand is permanently lost to the beach system and has a present market value in excess of $40 million. Section 111 of the 1968 Water Resources Development Act was written in response to concerns over erosion caused by the construction of similar entrances and authorizes "shoreline mitigation" projects.

Erosion to the south of this entrance in the Town of Palm Beach resulted in the construction of a sand transfer plant in 1958 by state and local interests and in substantial beach nourishment (shown in Figure 2-6.) This facility pumps a sand and water mixture through a pipe under the entrance. Over the years, this pumping facility has been enlarged and upgraded to be more effective and the Corps has studied further modifications and possible replacement. As of 1985, a total of 2.7 million cubic yards had been bypassed under the entrance by the sand transfer plant. In addition, by 1990, an additional 4.9 million cubic yards had been placed on the downdrift beaches by beach nourishment and sand dredged from the entrance channel (Dean, 1995). In 1986, the State of Florida, in recognition of navigational entrances on the stability of the state's shorelines, passed a law (Florida State Statute 161.141) with the goal of requiring maintenance of the coastal sediment transport facilities.

Case Study: Ocean City Inlet. This entrance was formed by a hurricane in August 1933, thus isolating the present city of Ocean City, Maryland, from modern-day Assateague Island. At this time, the Corps was studying the possibility of cutting an entrance approximately 8 miles south of the inlet's present location. Following the formation of the new inlet, the Corps immediately commenced construction of stabilizing jetties. As at many places along the nation's shoreline, this inlet, along with its jetties interfered

FIGURE 2-6 Port of Palm Beach, Lake Worth Entrance sand transfer plant. Photo courtesy of Robert Dean, University of Florida.

with the natural net southerly longshore sand transport and commenced eroding Assateague Island. Prior to the inlet formation, the City of Ocean City extended some distance south of its present location. However, the rapid erosion rates (approximately 15 feet per year; Dean and Perlin, 1977) resulted in this area being literally uninhabitable. The shoreline offset resulting from this inlet is shown in Figure 2-7. The shoreline to the north has advanced some 800 feet, and the island to the south has retreated such that at present its outer shoreline is farther landward than the inner shoreline was at the time that the inlet was cut. Assateague Island was incorporated into the National Seashore Program of the National Park Service in 1965. The Corps and the National Park Service are presently planning to nourish the northern portions of Assateague Island.

Case Study: The Gulf Intracoastal Waterway. The Gulf Intra-coastal Waterway (GIWW) is the nation's third busiest waterway. This dredged waterway parallels the Gulf of Mexico shoreline between Brownsville, Texas, and St. Marks, Florida, and is maintained by the Corps at a minimum

FIGURE 2-7 Ocean City inlet, Maryland. Photo courtesy of Robert Dean, University of Florida.

bottom width of 125 feet and a minimum depth of 12 feet. The first segment of the waterway in Texas was a canal that connected Galveston Bay and the Brazos River around 1853. The completed waterway etches its way through 423 miles of Texas coastal bays and floodplains (Texas Department of Transportation, 2000). Several studies have documented the effects of dredging the GIWW on turbidity and salinity in the 200-km-long Laguna Madre, which contains 75 percent of the seagrass habitats in Texas (Figure 2-8). Until the mid-1990s, dredged spoil from maintenance dredging was generally deposited in open water next to the GIWW channel. Between the mid-1960s and 1988, the net loss of more than 150 km^2 of seagrass cover in the lower Laguna Madre was attributed to reduced water clarity caused by maintenance dredging, and the species composition of the seagrass beds changed over another 190 km^2 due to salinity changes associated with changed circulation patterns caused by the GIWW channel, which connected the upper and lower Laguna Madre in 1949 (Onuf, 1994; Quammen and Onuf, 1993). In 1994, the National Audubon Society in

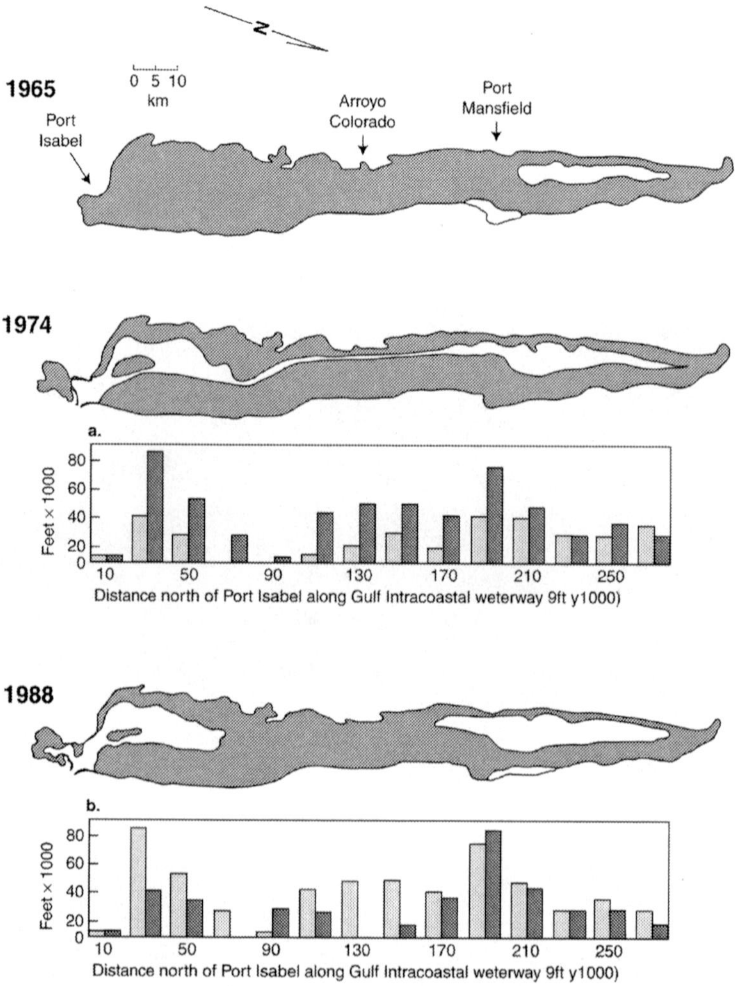

FIGURE 2-8 The timing and intensity of dredging activity in relation to seagrass distribution in the lower Laguna Madre of Texas. Seagrass is shown in gray, and bare areas are shown in white. Histograms show the length of channel dredged per 20,000 foot reach of the Gulf Intracoastal Waterway in the five years prior to seagrass surveys of (a) 1965 and 1974, and (b) 1976 and 1988. Some reaches require dredging every couple of years (reprinted from *Coastal and Shelf Science*, 39, C.P. Onuf, "Seagrasses, dredging and light in Laguna Madres, Texas," 75-91[1994], with permission of Elsevier Ltd., Oxford, United Kingdom).

conjunction with six other environmental organizations filed a lawsuit to stop the disposal of material dredged from the GIWW in the Laguna Madre between Corpus Christi Bay and Port Isabel. A settlement was reached wherein the Corps agreed to update the 1975 EIS prepared for the dredging of the GIWW project and to develop a long-term dredged material disposal plan in cooperation with the state sponsor, the Texas Department of Transportation.

Case Study: The Mississippi River Gulf Outlet Project. The Mississippi River Gulf Outlet (MRGO) project in southeast Louisiana was approved by Congress in 1958 and completed in 1965. The 76-mile channel was authorized to a depth of 36 feet and a surface width of 650 feet. The MRGO was dredged across the shallow waters of Chandeleur Sound and the marshes of St. Bernard Parish, which lie southeast of the city of New Orleans. The marshes and bald cypress trees in the wetlands in the coastal systems bisected by the canal declined rapidly as a result of the increased waves, tidal influence, and salinity (Figure 2-9). Since it was originally dredged, the canal has widened threefold in some places. The Corps' New Orleans District estimates that the conversion of land to open water in the area immediately adjacent to the canal is nearly 3,400 acres (Figure 2-10). The canal has contributed to the loss or severe degradation of approximately 10,300 acres of brackish marsh, 4,200 acres of saline marsh, and 1,500 acres of cypress swamps and levee forest. In addition, the canal has caused more than 36,000 acres of marshes as well as Lake Pontchartrain to become more brackish (Louisiana Department of Natural Resources, 2002). The Corps is presently studying the possibility of closing the MRGO to ship traffic.

LINKAGES BETWEEN RIVER BASINS AND COASTAL SYSTEMS

Terrestrial watersheds supply water, sediment, and nutrients to the coastal zone. Disruption in this flux, primarily from reservoir construction, impacts the physical and ecological characteristics of coastal systems. The solution to coastal problems may often be found far upstream, illustrating

FIGURE 2-9 Image of a bald cypress swamp impacted by saltwater intrusion near New Orleans, Louisiana, resulting from the building of the Mississippi River Gulf Outlet, a 76-mile-long, man-made navigational channel connecting the Gulf of Mexico to the city of New Orleans. The U.S. Army Corps of Engineers (1999c) report speculates that 1,500 acres of cypress swamps and levee forests have been destroyed or severely altered as a result of the MRGO. In addition, the Corps (1999c) report estimates the loss of 3,400 acres of fresh or intermediate marsh, 10,300 acres of brackish marsh, and 4,200 acres of saline marsh. Photo courtesy of Rex H. Caffey, Louisiana State University.

the importance of integrated water systems planning and the significant multi-jurisdictional challenges faced when managing the coastal zone. Terrestrial impacts on the coastal zone can be seen over a wide range of scales, from water quality and shoreline loss in the Gulf of Mexico, associated with agricultural practice and river regulation. in the Mississippi-Missouri-Ohio River system, to beach erosion and harbor siltation in many smaller coastal watersheds.

Hypoxia and eutrophication in many of the nation's coastal waters illustrate the influence of the delivery and quality of inland surface waters

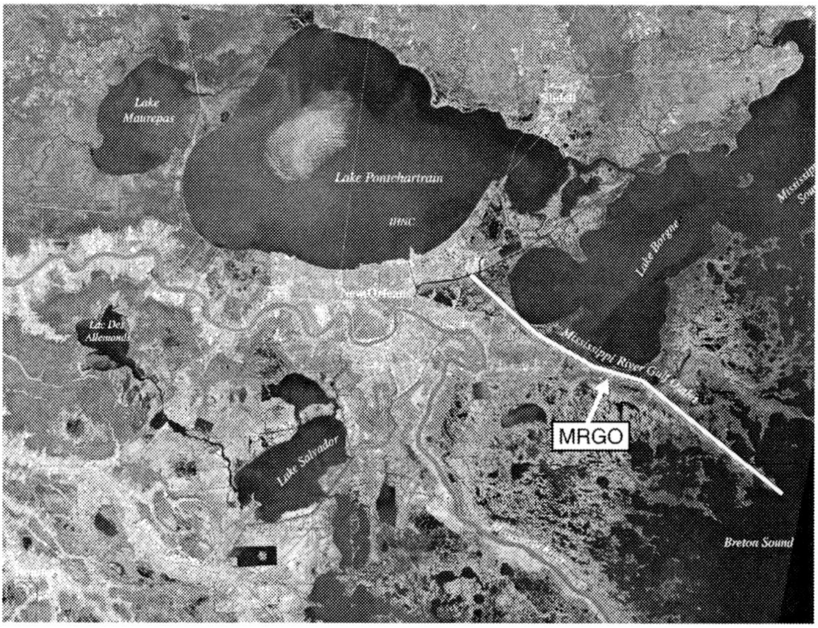

FIGURE 2-10 Image of the 76-mile-long Mississippi River Gulf Outlet, Lake Borgne, and adjacent coastal wetlands. New Orleans urban area located in upper left, adjacent to Lake Pontchartrain. Figure courtesy of John Barras, U.S. Geological Survey, Lafayette, Louisiana.

on environmental quality in the coastal zone. Several investigations during the past 10 years have linked chronic bottom-water hypoxia (<2 mg/L dissolved oxygen) in the north-central Gulf of Mexico to increased nutrient loading in the Mississippi River (Bratkovich et al., 1994; Dortch et al., 1994; Justic et al., 1993; Rabalais and Turner et al., 2001; Rabalais et al., 2002). In midsummer, the aerial extent of bottom-water hypoxia may cover up to 20,000 km^2, which concerns fishery managers and others interested in estuarine and coastal water quality. Hypoxia is also common in coastal bays and estuaries and is often linked with activities in adjacent watersheds. Approximately half of the 58 estuaries within the U.S. Gulf of Mexico coastal region have evidence of nitrogen and phosphorus loading from upland sources at rates that indicate eutrophic or near-eutrophic conditions. Nutrients from adjacent watersheds have been identified as a major cause of water quality degradation in coastal Florida, Louisiana, and Mississippi (NRC, 2000; USEPA, 1999).

Suspended sediment loads carried by the Mississippi River have decreased by one-half since European immigrants settled the Mississippi Valley. Most of the decrease in sediment load has occurred since 1950, because the largest sources of natural sediment supply were cut off from the main stem of the Mississippi River by the construction of large reservoirs on the Missouri and Arkansas Rivers (Meade, 1995). This decrease in riverborne sediments has adversely affected coastal habitat integrity in the Mississippi River delta region, where roughly 1 million acres of coastal wetlands have been lost since 1940. The decrease in suspended sediments and the channeling of sediments into deep water to facilitate navigation have contributed to significant land loss in coastal Louisiana. Proposed coastal protection and restoration projects to address this problem in coastal Louisiana may cost as much as $14 billion (Committee on the Future of Coastal Louisiana, 2002).

The trapping of sediment in reservoirs constructed for flood control and water supply can reduce the supply of sediment needed to replenish coastal beaches. Efforts to mitigate the loss of sediment from coastal systems are under way, although few projects have been successfully completed. Difficulties include the cost of sediment removal, the long distances from the reservoirs to the coastline, and the potential for contamination in the deposited sediment and its interstitial waters. Consideration is being given to the removal of some deteriorated dams, which should allow some of the trapped sediments to be transported to the coastline.

Land disturbance and erosion in terrestrial watersheds can also increase the supply of fine sediments to coastal waters requiring dredging of navigable coastal waters, and delivering nutrients that degrade the water quality of estuaries. It has been recognized that the recovery of the Chesapeake Bay requires a reduction in sediment and nutrient supply from terrestrial sources (Chesapeake Bay Program, 2000).

Projects that alter the structure or function of one component of a natural water resources system will tend to impact not only the immediate project area but other parts of the system as well. Appropriate evaluation of project impacts requires an understanding of the broader system within which the project is placed. If the cumulative effects of reservoir and levee construction on Mississippi River sediment transport had been analyzed as part of the Mississippi River and Tributaries Project, wetland losses in the Mississippi deltaic plain could have been foreseen, allowing for the possibility of earlier and more effective mitigation. A system perspective is needed not only for impact assessment, but also for successful project design. Elements of a river basin or coastal system and its history (as well as other existing or proposed projects) can have important influences on the success of a water resources project.

SPATIAL AND TEMPORAL SCALES FOR PLANNING AND MANAGEMENT OF WATER RESOURCES PROJECTS

Integrated water systems planning and management require a balanced evaluation of all relevant factors over appropriate scales of space and time, such that multiple objectives can be investigated, unintended consequences can be identified, and the complete costs and benefits of a project may be evaluated. Such an approach may often require that project planning and evaluation be conducted at spatial and temporal scales larger than specifically required for isolated individual projects. Because of the range of factors involved in evaluating modern water projects, it may often be necessary to evaluate different factors at different spatial and temporal scales.

Spatial Scales for Water Resources Project Planning

As some of the preceding case studies illustrate, water project activities can affect, for better or worse, hydrologic, ecological, and economic conditions beyond the immediate project area. Flood protection levees in one location, or the progressive construction of many levees, can exacerbate flooding in downstream portions of a waterway. Project alteration of hydrological regimes and sediment transport directly affect the magnitude and timing of nutrient and contaminant loadings to downstream systems. Construction of dams and water diversion projects and the dredging of ports affect water chemistry and quality, including water temperature and nutrient loading. Projects thus potentially alter the growth and reproduction of plants and animals, the location of fisheries, and the intensity and type of local commerce and shipping. Corps permitting and projects in rivers and coastal systems also potentially disrupt movements of animals, which in turn can affect the sustainability of fisheries and the trophic structure of aquatic systems. Development or maintenance of navigation channels will influence the relative costs of different shipping alternatives, thereby potentially impacting a wide range of economic activities. Reoperation of storage reservoirs for ecosystem or recreation objectives can influence the timing and amount of hydroelectric generation, which can have repercussions throughout a regional power grid.

The complexities of defining the appropriate spatial scale are particularly difficult in the context of environmental stewardship, which requires consideration of the ecosystem affected by the project. Among the basic ecosystem principles noted in *Ecosystem-Based Fishery Management*, a National Marine Fisheries Service (NMFS, 1998) report to Congress, are the principles that "multiple scales interact within and among ecosystems," that "components of ecosystems are linked," and that "ecosystem boundaries are open." Ecological systems are connected by, among other things, water flow, the movement of people, and the activities of plants (e.g., photosynthesis; the influence on transport and transformation of nutrients) and animals inhabiting the ecosystems. All of these vectors influence the transport of materials, which may indirectly affect an ecosystem's natural balance within its watershed or coastal system. Water flow transports nutrients, sediments, and contaminants from the landscape to waterways and ultimately downstream to estuaries and the ocean. Many organisms, including economically and ecologically important species and their prey, are dependent on large portions of watersheds to complete their life cycle.

Almost without exception, the watershed approach is endorsed by scientists, water planning experts, and the National Research Council (NRC, 1992, 1996, 2000, 2001a, 2002a) as the framework in which the physical and ecological aspects of water resources planning should occur. In the context of coastal systems, a comparable framework would be defined in terms of the region encompassing significant flows of water, sediment, nutrients, and contaminants. Reasons supporting watershed and coastal systems planning include the inherent connectedness of hydrologic and ecological systems; the importance of location within the landscape to both habitat function and the impact of anthropogenic activities; and the localization of potential cumulative effects and unintended consequences within a watershed or coastal system.

In the terrestrial realm, a framework for water project evaluation is given by the hydrological unit hierarchy shown in Figure 2-2. Used as a basis for project evaluation, this system makes clear the nested nature of potential project benefits and costs and would provide a consistent basis for project analysis. Use of hydrologic unit codes and "nested watersheds" is supported by numerous studies (Seaber et al., 1987; U.S. Department of Agriculture, 2003). The analogous spatial planning unit for coastal waters is the "estuarial region" and the "coastal unit" (NRC, 1999b).

Regardless of the basic heuristic for determining the spatial boundaries on possible project effects, a set of clearly defined temporal and spatial scales for water planning investigations cannot be specifically prescribed for all cases. The hydrologic, ecologic, and economic setting of any particular water resource project will vary from that of other projects. While one might ideally envision a national analysis for every Corps water project to ensure consistency and fit with a national water policy or set of objectives, such an approach would be financially impractical. Nonetheless, the essential starting point for evaluating any water project is the identification of the essential factors and objectives of that project and their relevant spatial scales. There may be instances in which a local focus is sufficient, but there are also cases (e.g., commercial navigation on navigable waterways, waterways that provide critical habitat for migratory endangered birds, projects that contribute significantly to regional electric power grids) where project planning and evaluations must extend beyond watershed boundaries, at least for some essential aspects of the project. Because the spatial scale for project evaluation varies from case to case, flexibility in defining the appropriate spatial scale is essential for effective, integrated water resources management. Current cost-sharing and project time lines can pressure the Corps to limit the general scope and spatial scale of a project evaluation to an area in the immediate vicinity of the proposed project. Revisions to project planning arrangements that would help address these pressures are discussed in Chapter 5.

Temporal Scales for Water Resources Project Planning

The processes that operate within watershed and coastal systems operate at varying rates. Thus, the magnitude of any measurable parameter that may be important to the design or operation of a water resource project is not fixed and varies through time. This temporal variability has been recognized in engineering practice for decades and has historical been addressed by assuming that important design variables will vary within a predictable range over a given time, allowing designers to build in a margin of safety. However, as the objectives of water resource projects have increased and as efforts to more fully account for the behavior of complex natural systems are made, sensitivity to temporal variation and the scale at which these variations takes place has become increasingly important. For water projects, the most significant factors whose time variability must be accounted for are those associated with climate.

Consideration of Climate Variability and Change in Water Resources Planning

A large body of research relating to climate variability has been published in water resources planning and management archival literature during the past 15 years (Cayan et al., 1998, 1999; Kahya and Dracup, 1993). This research indicates that there have been significant variations in interannual and interdecadal precipitation and streamflow in selected areas of the United States, most notably in the Pacific Northwest, the Pacific Southwest, and the southeastern United States (Cayan et al., 1999). The causes of these variations have been determined, allowing for prediction with lead times of up to seven to nine months (Goddard et al., 2001). It also has been demonstrated that the incorporation of predictable interannual and interdecadal streamflow variations in the planning and management of water resource systems can lead to considerable increases in hydroelectric and water supply revenues, flood reductions, and better approaches to drought management (Hamlet and Lettenmaier, 1999; Piechota and Dracup, 1996).

Longer-term trends in temperature and precipitation are also well documented. Instrumental records indicate that average annual precipitation and temperature increased during the twentieth century in the United States and over most of the northern hemisphere (IPCC, 2001; Melillo et al., 2000; NRC, 2001b). These changes were most pronounced at high latitudes. Alaska has warmed an average of 2° C since 1950, while the lower 48 states warmed roughly 0.6° C during the past 100 years (which is roughly the global average). Although there is a great deal of regional and local variation, average annual precipitation increased during the twentieth century by approximately 5 to 10 percent in the conterminous United States, with much of it due to an increase in the frequency and intensity of heavy rainfall (Karl and Knight, 1998). Climatic changes of this nature have practical significance to water resource managers and numerous implications for the design of water resources projects.

Accelerated sea-level rise is regarded as one of the most costly and most certain consequences of global warming. If sea-level rise increases at rates projected by the United Nation's IPCC (2001) during the next century, many of the world's low-lying coastal zones and river deltas could be inundated. Along the low-lying southeastern coastal margin, where the majority of U.S. flood losses occur in an average year, effective design and maintenance of flood control works will require the incorporation of long-term sea-level and precipitation trends in project design.

Case Study: The California Reservoir System. California water management is charged with meeting many socioeconomic challenges including providing a reliable water supply; protecting developed areas from flooding; and balancing competing water uses related to urban, agricultural, and environmental benefits. California's hydrology is extremely variable, involving multi-year droughts and short-duration extreme floods, and this variability increases the difficulty of water resources management. Regional water systems have been constructed to dampen the effects of droughts and floods on beneficial use. Two of the major regional systems serving California are the Central Valley Project (CVP) and State Water Project (SWP) systems, each collecting surface water runoff from the northern half of the state for redistribution throughout the state.

Operation of the CVP-SWP system is driven by risk management associated with floods and droughts. Management guidelines are not translated into detailed risk assessments but are expressed through general operating principles (Brekke, 2002). These principles include the following: (1) seasonal to annual CVP-SWP operation requires hydrologic anticipation of supply availability and flood potential; (2) information supporting this anticipation comes from past experience (i.e., climatology) and snow survey data collected during winter; (3) the consequence of using supportive information errantly (i.e., over anticipation of supply availability or underestimation of flood potential) is a system operational state where drought and flood vulnerability is increased (without identifying specific consequences of that increased vulnerability); and therefore (4), the management action is to avoid using information errantly by using it very *conservatively* (e.g., 90th percentile exceedance anticipation of supply availability). Past results of operations have been mostly satisfactory; however, the periodic occurrence of floods (e.g., New Years 1997 in the Sacramento Valley) and drought-induced water shortages (e.g., 1987-1993), coupled with the increasing competition for California water, leads to the question of how to conduct hydrologic anticipation more aggressively while maintaining adequate risk protection and providing a more flexible supply management capability.

Improvements in longer-term water management can look to climate-weather connections between Pacific basin phenomena (e.g., El Niño Southern Oscillation [ENSO]) and subsequent western U.S. hydroclimatic variations (i.e., weather and hydrologic variations). For example, it is widely understood that when ENSO is in its El Niño phase during summer months, there is likely to be increased winter precipitation in the Pacific Southwest (PSW) and decreased winter precipitation in the Pacific Northwest (PNW). Although ENSO has a relatively weak connection to

Northern California hydroclimatic variations compared to the PNW and PSW, this illustration makes the point that distant climatic information in the Pacific can be monitored and would seem to improve our foresight of hydrologic conditions and affect how we manage our water resources. Despite the available data, distant climatic information is seldom used to influence large regional system operations.

Planning for Continued Climate Change

The changes in climate that are projected to occur during the next century could have a significant impact on the planning, design, and operation of water systems in the United States. Particularly, climate change could have a significant impact on river basin deltas and coastal regions. In spite of current levels of uncertainty associated with General Circulation Models (GCMs) that are used to simulate and predict climatic change, there is general agreement among climate scientists that (1) the amount of greenhouse gases such as carbon dioxide is increasing in the atmosphere; (2) the world wide atmospheric temperature increased approximately 0.5° C (0.6 ± 0.2° C) during the twentieth century; and (3) the average global sea-level rise ranged from 20 to 65 mm from 1910 to 1990. Furthermore, as stated in an assessment of the Intergovernmental Panel on Climate Change, "globally it is very likely that the 1990s was the warmest decade, and 1998 was the warmest year" in the instrumental record (1861-2000) (IPCC, 2001, p. 2).

The three parameters of global climate change that will likely have the greatest impact on water resources in the United States are air temperature changes, sea-level rise, and precipitation.

1. *Temperature.* Based on predictions from nine GCMs, the IPCC reports a potential air temperature increase during the next 100 years ranging from a low of 1.0 percent to a high of 5.2 percent by the year 2100 over a base of 1961 to 1990 values (IPCC, 2001, p. 541).
2. *Sea level.* Based on the predicted sea-level rise from 14 GCMs, the IPCC assessment reports a potential global 0.10- to 0.87-meter increase in sea level from 1990 to 2100 (IPCC, 2001, pp. 670-671).
3. *Precipitation.* Based on predictions from nine GCMs, the IPCC reports a potential precipitation increase ranging from a low of 1.0 to a high of 8.9 percent by the year 2100 over a base of 1961 to 1990 values (IPCC, 2001, p. 541).

The impact of global climate change (based on increases in air temperature, sea-level rise, and precipitation) potentially could be a significant problem and have negative impacts on Corps planning and management of water resource projects. For example, hurricane protection levees in the New Orleans region are presently designed and maintained by the Corps to protect the city from storm surge flooding. Due to subsidence associated with groundwater withdrawals, levees along the Mississippi River that prevent sediment deposition, global sea-level rise, drainage and oxidation of organic soils, and natural compaction and dewatering of the deltaic sediments upon which the city has been constructed, most of the city of New Orleans lies below mean sea level (MSL). Hurricane protection levee design heights range from about 4.5 to 6 meters above MSL. The levees along the Lake Pontchartrain shoreline are designed at a height that exceeds the surge and waves of a Category 3 hurricane. The design for these levees assumes no increase in MSL (Burkett et al., 2003). The original congressional authorizations for these projects did not provide the Corps with the authority to include sea-level rise in its plan for maintaining these levees. The Corps is planning to review these projects, however, to determine if changes should be made to account for sea-level rise in future authorizations for these and other flood protection projects in the New Orleans region (Alfred C. Naomi, U.S. Army Corps of Engineers, New Orleans District, personal communication, December 4, 2003).

SUMMARY

The spatial and temporal complexities of river basin and coastal systems discussed in this chapter present significant challenges to effective water resource planning and implementation. Historically, the Corps has been at the forefront of engineering practice and has undertaken much of the pioneering work in relevant civil engineering fields. The Corps has long embraced the underlying philosophies of systems planning and watershed approaches to project implementation and management. As growing awareness of the importance of coastal systems emerged, the Corps has attempted to incorporate greater understanding of these systems in water resource planning in the nation's coastal regions. However, as the nation's expectation for such projects has changed and greater emphasis has been placed on maintaining or restoring ecological function in the nation's river basin and coastal systems, the portfolio of water resource projects planned and implemented by the Corps has changed. As a consequence, uniform application of key systems planning concepts in its water resources planning and implementation efforts is a goal the Corps continues to pursue. The following chapters explore some of these concepts in greater detail and attempt to recommend how these goals may be achieved.

3

Role of the U.S. Army Corps of Engineers in Environmental Restoration and Stewardship

Chapter Highlights

The Corps' commitment to ecosystem restoration and environmental stewardship increases the demand for and complexity of integrated water project planning and evaluation. Cumulative impacts of existing and planned projects must be considered, and effective methods for evaluating environmental as well as economic objectives must be implemented. This chapter examines the Corps' obligations in environmental stewardship in its role as a water project agency, as well as an environmental regulator.

Given the large number and wide range of existing and proposed projects, the Corps plays a dominant role in the stewardship of the nation's ecosystems. The Corps has increasingly embraced this role and now states that environmental stewardship is a central part of its mission (U.S. Senate, Committee on Environment and Public Works, 2002). Environmental restoration is a significant and growing part of the Corps' portfolio. Evaluation of environmental impacts and benefits of project alternatives are now a standard part of Corps practice, although the methodology is neither fully developed nor consistently implemented. Chapter 2 illustrates the complex interconnectedness of hydrologic systems and the ecosystems they support. Successful environmental restoration projects, like more traditional water resources projects, require integrated systems planning. The unique challenges and opportunities presented by this increased focus on restoration and stewardship thus warrant fuller exploration.

ECOSYSTEM RESTORATION IN RIVER BASINS AND COASTAL SYSTEMS

In the early 1990s, ecosystem restoration was formally stated as a primary mission of the Corps of Engineers civil works program. The Corps' objective in ecosystem restoration planning is to "contribute to National Ecosystem Restoration (NER). These contributions, or NER outputs, are defined in the Corps *Planning Guidance Notebook* (*PGN*) as "increases in the net quantity and/or quality of desired ecosystem resources" (U.S. Army Corps of Engineers, 2000b, Section 2.2[b]).

When Lt. General Henry Hatch, former chief of engineers and commander of the U.S. Army Corps of Engineers, addressed the American Society of Civil Engineers in the fall of 1989, he said, "It is we engineers who hold most of the keys to the solutions of the world's environmental problems." As Corps leaders accepted their new mission to protect the environment, they also conceded that the Corps' past practices unintentionally damaged sensitive ecosystems and asserted that adequate engineering expertise exists to correct these problems. The $9.4 billion Corps budget for fiscal year (FY) 1990-1991 included roughly $1 billion for environmental restoration projects, which ranged from hazardous waste cleanups at military bases to the creation of wetlands. The FY 1990-1991 budget also included funds to modify several existing Corps projects for the purposes of ecosystem restoration, such as restoring wildlife areas along the Tennessee-Tombigbee Waterway and rebuilding fish habitat along the Columbia River. Funding for the Corps' environmental programs has increased dramatically over the past decade (Figure 3-1); most notable is funding for extraordinarily large-scale projects such as the Everglades restoration and the restoration activities being considered in Louisiana.

By 2003, the Corps had proposed environmental enhancement and restoration projects in 35 states; some examples include requests for $95 million for fish habitat restoration in the Columbia River basin, $22 million for Missouri River fish and wildlife mitigation, and $127 million for ecosystem restoration in South Florida. Environmental enhancement and restoration projects comprise one-third of new Corps projects (including

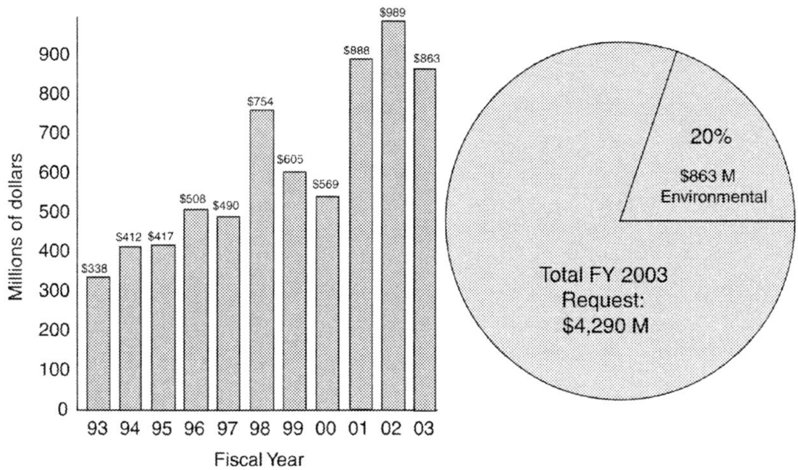

FIGURE 3-1 Increase in the environmental portion of the Corps' budget over the past few decades. Figure courtesy of U.S. Army Corps of Engineers.

reconnaissance "surveys") proposed for FY 2004 (U.S. Department of the Army, 2003). With the approval of the $8 billion Everglades ecosystem restoration program in the Water Resources Development Act of 2000 and the ongoing discussion of devoting approximately $14 billion to a Louisiana coastal area restoration project, the Corps has embarked on a large-scale "ecosystem restoration" approach. With this new approach, the Corps is undertaking restoration as a primary project objective, without dependence on a flood control or beach nourishment project as a basis for plan initiation.

For environmental restoration to be a successful component of Corps activities, it is important to acknowledge it is still a young science. Consequently, it is critical that the Corps allocate sufficient funds to monitor the efficacy of restoration efforts and generate information that will ultimately improve the success of restoration projects (this point is discussed in the newly released [NRC, 2004a] report *Adaptive Management for Water Resources Project Planning*). The duration, scope, and scale of monitoring of these projects should be sufficient to predict successful restoration and mitigation efforts. Designing and monitoring restoration projects so that either success or failure of a project can provide information to be used in improving the effectiveness of future efforts is an economically and environmentally sound practice. Currently, the Corps is limited to spending only 1 percent (up to 3 percent for adaptive management projects) of project funds on monitoring and assessment for restoration projects. *Adaptive Management for Water Resources Project Planning* specifically called on Congress to provide new authority and direction to strengthen the Corps' ability to adjust operations of existing projects (in order to increase overall project benefits) and increase its ability to monitor post-construction changes. Furthermore, the report called for Congress to appropriate sufficient funding (with appropriate cost-sharing by cosponsors) for post-construction monitoring and evaluation of environmental and economic objectives and subsequent outcomes (NRC, 2004a). Nowhere is this need more acute than in large, complex, environmental restoration projects.

Maximizing the probability of current and future success in environmental restoration in complex systems with multiple jurisdictions requires the Corps and its partners to adopt a consistent approach to adaptive management. Accepted adaptive management approaches argue that once a project is designed and constructed, a plan is put in place to implement changes in the management of restored habitat in response to undesired system changes. A successful example of this approach can be seen in the Delaware Bay, where coastal restoration is being conducted by the New Jersey-based Public Service Electric and Gas Company (PSE&G) (Mitsch and Jørgensen, 2003, 2004; Weinstein et al., 1997, 2001). This restoration uses the extent of coverage of coastal marsh grass (*Spartina alterniflora*) and other desirable flora to trigger management responses. The Corps is also adopting this management approach in other restoration efforts such as the new plan for oyster restoration in the Virginia portion of Chesapeake Bay (see Box 3-1).

Role of Ecological Engineering

The emerging discipline of ecological engineering can greatly aid the Corps in this environmental-steward role and in the rehabilitation and restoration of habitats. Sustainable ecosystem restoration requires emphasizing entire ecosystem function over a species-by-species analysis. Ecosystems are inherently self-designing, and restoration efforts should recognize this process through the application of sound environmental engineering principles. Traditionally, restoration efforts have focused on improving land and water resources for specific plant and animal species and have not taken a holistic approach to planning and management. Box 3-2 provides some background on an ecological engineering framework and how the Corps may benefit from planning within that framework.

CUMULATIVE EFFECTS OF THE CORPS' ACTIVITIES IN RIVER BASINS AND COASTAL SYSTEMS

A central concept of ecosystem analysis is that the response of individual species—and the ecosystem as a whole—to any event cannot be examined in isolation. The cumulative effects of a variety of anthropogenic stressors—not simply individual stressors or incremental changes from human influence (Breitburg and Riedel, in press)—must be examined. The potential for many small individual projects to lead to large-scale habitat degradation, the occurrence of interactions among stressors, and the importance of threshold values of environmental variables to organisms make it apparent that a narrow evaluation of individual stressors or incremental changes can severely underestimate the potential for environmental damage.

Instances of stressor interactions and threshold effects are common. For example, the expression of trace element toxicity can depend on nutrient loadings and ratios (Breitburg et al., 1999); habitat alteration can influence the success of invasive species (Stachowicz et al., 2003); and small changes in dissolved oxygen (affected by nutrient loadings and flow) can increase the mortality of aquatic organisms (Miller et al., 2002). Although single events or large-scale stressors can negatively impact a system, over time unchecked small or interactive occurrences can be equally, if not more, destructive. Alternatively, an incremental increase in habitat alteration that affects a particular species' population or the overall ecosystem function may be small, yet the cumulative impacts can be great.

A piecemeal approach to environmental management without strong focus on cumulative effects is likely to result in a gradual erosion of habitat quality and quantity. A change in procedures for project and permit evaluation is needed so that cumulative effects can be factored in when estimating environmental impacts. For example, the potential impact of increasing the depth of a shipping channel should be evaluated not only by estimating the impact of the new increment to be dredged, but also by estimating the total impact of a shipping channel from the proposed dredging.

Historically, the Corps estimated ecological impacts of an incremental change when a project was expanded. For example, the environmental impact statement (EIS) examining the effects of extending several locks on the Upper Mississippi River-Illinois Waterway considered only the environmental risk resulting from increased barge traffic, not the long-term effects of operation and maintenance of the entire navigation system (NRC, 2001a).

BOX 3-1
Oyster Restoration Plan for the Virginia Portion of Chesapeake Bay

The population and therefore the harvests of the Eastern oyster (Crassostea virginica) have declined during the past century due to a combination of overharvesting, habitat destruction, and disease (Mann, 2000; Rothschild et al., 1994). In response, oyster restoration efforts have been under way in both the Maryland and the Virginia portions of the Chesapeake Bay.

The 2003 plan for oyster restoration in the Virginia portion of Chesapeake Bay exemplifies the value of many of the recommendations of this panel, which are also embodied in the U.S. Army Corps of Engineers' Environmental Operating Principles. These include the following:

- making the best use of available scientific expertise;
- taking an innovative approach to comparing potential economic and environmental benefits of a project
- strongly incorporating adaptive management into the plan to acknowledge the uncertainty involved in successfully restoring oysters; and
- increasing funding for monitoring to evaluate project success, allow adaptive management strategies to be implemented, and plan the project on a watershed scale (D. Schulte, Virginia Sea Grant, personal communication, 2003).

The oyster restoration plan was developed with close collaboration of and advice from scientists at the Virginia Institute of Marine Sciences. This allowed the plan to be developed using the most up-to-date scientific information. The Corps adopted the consensus recommendation on the strategy for oyster restoration with the highest likelihood of success—a plan that included consideration of oyster genetics, disease susceptibility, and characteristics of the physical environment. State-of-the-art approaches will be used to evaluate and improve the success of Chesapeake Bay restoration projects, including tracking genetic markers to provide information on the contribution of planted

continued

BOX 3-1 Continued

oysters to local populations (a measure of the success of the project) and using newly developed disease-tolerant oyster strains.

The planning process also broke new ground in that it included a critical examination of two categories of restoration goals by comparing the value of constructed oyster harvest grounds with the value of creating sanctuaries in which oyster removal would be prohibited. The Corps decided to focus all of its efforts on the creation of sanctuary reefs because the benefits associated with the ecological function of oysters would be lost if harvesting were permitted, and these ecological benefits were not compensated by the economic benefits associated with the creation of harvest sites. Although the adopted plan deviates from that originally proposed by the state sponsor, over time the predicted benefits to the fishery through increased oyster reproduction and seed production are greater than would be expected if the majority of effort had been designated for construction of harvest sites.

The oyster restoration plan also emphasizes adaptive management and deviates from standard Corps practice by designating 10 percent of the budget for monitoring and adaptive management rather than the usual 1 to 3 percent. The lower, more typical level of funding for monitoring was judged as inadequate to address the uncertainties surrounding oyster restoration. More extensive monitoring will permit a more accurate assessment of the success and benefits of the restoration projects and will provide data needed for the implementation of corrective actions should they be required.

Finally, project planning included spatial scales far larger than the size of individual restoration projects and considered the interconnectedness of the aquatic environment. Planning has been on a state-wide basis and includes the entire Virginia portion of the Chesapeake Bay and its tributaries, and 30 monitoring sites throughout the region.

Increased attention to the cumulative effects of Corps projects and permitting is also important because river basins and coastal systems are exposed to multiple stressors resulting from both human activities and natural disturbances such as storms and floods. Altering the landscape or hydrology potentially increases the susceptibility of ecosystems and indi-

BOX 3-2
Ecological Engineering

Although there was some early use of the term ecological engineering in the 1960s, the use of this term and the development of the ecological engineering field in industrialized countries has occurred more recently and did so primarily due to the following factors: (1) the loss of confidence that all environmental problems can be solved through conventional technological means; (2) the continued loss of sustainable ecosystems including rain forests, coral reefs, riverine habitat, riparian forests, and wetlands; and (3) the recognition that many solutions to environmental problems and pollutants merely shift these problems and pollutants from one form or location to another (Mitsch and Jørgensen, 2003, 2004).

Ecological engineering involves both (1) the restoration of ecosystems that have been substantially disturbed by human activities such as environmental pollution or land disturbance, (2) the development of new sustainable ecosystems that have both human and ecological value (Mitsch and Jørgensen, 2004). Five ecological engineering principles have been suggested by Mitsch and Jørgensen (2003, 2004):

1. It is based on the self-designing capacity of ecosystems.
2. It can be the acid test of ecological theories.
3. It relies on system approaches.
4. It conserves non-renewable energy sources.
5. It supports biological conservation.

Several restoration fields have developed somewhat independently of ecological engineering, and many appear to have the design of ecosystems as their theme as well. A definition of ecological restoration, which resulted from a National Academy of Sciences study in the early 1990s on aquatic ecosystem restoration, is "the return of an ecosystem to a close approximation of its condition prior to disturbance" (NRC, 1992, p. 5). Although related to ecological engineering or even a part of it, many restoration approaches seem to lack one of the two important cornerstones of ecological engineering, namely: (1) recognizing the self-designing ability of ecosystems or (2) basing the approaches on a theoretical foundation in ecology, not just empiricism. Furthermore, restoration of ecosystems to what they

continued

BOX 3-2 Continued

were previously thought to be is not always possible due to changed hydrologic, geomorphic, and biogeochemical factors. A popular view is that ecological restoration is a subfield of ecological engineering. In any event, the fields of ecological engineering and ecosystem restoration, slightly different in their goals, yet conjoined because they both involve designing ecosystems, are distinctly different from the better-known fields of environmental engineering, industrial ecology, and biotechnology (see discussion in Mitsch and Jørgensen, 2004). Ecology as a science is not routinely integrated with engineering, even in environmental engineering programs. Creating stronger linkages between these fields will be a challenge and an opportunity for organizations such as the Corps as these organizations carry out complex restoration projects.

vidual organisms to the negative effects of additional stressors. In some cases, two or more stressors interact in ways that increase the environmental problems or introduce new challenges. That is, although each stressor may have little or no effect on the system by itself, when stressors occur simultaneously the overall impact can be greater than the sum of the individual effects. One obvious example of cumulative effects is well recognized in planning water resource projects dealing with flood prevention. The construction of a single levee or floodwall has limited impact on the overall system because floodwaters are simply diverted away from protected areas into adjacent floodplain or downstream. However, extensive construction of levees and floodwalls within a river basin results in higher flood stages downstream. The Corps recognizes this potential for cumulative effects of flood control structures and accounts for it in project design. However, building complex water resource projects, significantly changing land uses, and extensively modifying geochemical or hydrologic cycles (through use of pesticides or fertilizers or water diversion) are not as easily accounted for by Corps project planners, especially given that the Corps is not involved in the majority of these activities. The consequences of these multiple stressor interactions may be expressed at a distance from the site of impact, affecting the success of restoration projects and increasing the temporal variability of important ecosystem processes (Breitburg and Riedel, in press).

UNDERSTANDING THE CORPS' REGULATORY RESPONSIBILITIES

In addition to its role in planning, developing, and maintaining water projects, the Corps also plays a major role in environmental stewardship through its regulatory responsibilities regarding the disposal of fill (sediment) in U.S. waters. For completeness, this section first describes the scope and responsibilities of the Corps.

The Corps of Engineers was granted the authority to regulate activities in U.S. navigable waters by the Rivers and Harbors Appropriation Act of 1899 (33 U.S.C. 401 et seq.). The regulatory responsibilities created by this act focused on controlling activities that would interfere with navigation. Section 9 of the act required Corps of Engineers approval of "any bridge, dam, dike, or causeway over or in any port, roadstead, haven, harbor, canal, navigable river, or other navigable water of the United States." Section 10 of the act states that "it shall not be lawful to excavate or fill, or in any manner to alter or modify the course, location, condition, or capacity of, any port, roadstead, haven, harbor, canal, lake, harbor of refuge, or enclosure within the limits of any breakwater, or of the channel of any navigable water of the United States" without prior permission of the Corps of Engineers. Section 10 permits include structures (e.g., piers, wharves, breakwaters, bulkheads, jetties, weirs, transmission lines) and work such as dredging or disposal of dredged material, or excavation, filling, or other modifications to navigable waters of the United States. "Navigable waters" are defined as waters that have been used in the past, are now used, or are susceptible to be used as a means to transport interstate or foreign commerce up to the head of navigation. The environmental consequences of these types of activities were acknowledged with the passage of the Federal Water Pollution Control Act Amendments of 1972; this law became commonly known as the Clean Water Act (CWA). The CWA broadened the Corps' authority over dredging and filling in "waters of the United States."

Section 404 of the CWA of 1972 prohibits the discharge of materials, such as dredged sediments, into coastal and inland waters of the United States unless authorized by a permit issued by the Corps of Engineers or a state with a regulatory program approved by the U.S. Environmental Protection Agency (USEPA). The waters of the United States, as defined by the Corps and the USEPA for the purposes of the Section 404 regulatory program, include most wetlands.

The Corps issues two types of permits under its regulatory programs: general permits and standard permits. General permits are issued for activities that have only minimal impacts on the aquatic environment. General permits can be issued on a nationwide, regional, or state basis. Examples of activities that are permissible under nationwide permits include aids to navigation, some forms of bank stabilization, installation of utility lines, Coast Guard-approved bridges, boat ramps, oil spill cleanup, and modification of existing marinas. In total, the Corps issued 35,768 nationwide permits under the Section 404 program in FY 2002 (see Table 3-1).

A regional permit may be issued by a district engineer when proposed activities are similar in nature and cause minimal environmental impact (both individually and cumulatively) and if the regional permit reduces duplication of regulatory control by state and federal agencies. Individual permits, the most common form of permit issued by the Corps, are issued on a case-by-case basis for projects involving the discharge of materials into waters of the United States when a review of the proposed activity is in the public interest, as it is defined in the Corps' regulatory program regulations that govern administration of the Section 404 permitting program of the CWA. Individual permits are issued following a full public interest review of an individual application for a Corps permit. After evaluating all comments and information received, a final decision on the application is made. The Corps may decide to deny the proposed activity, permit the proposed activity, or permit the proposed activity with modification. The Corps (or an approved state program) may attach conditions to permits, including a requirement for compensatory mitigation. The permit decision is generally based on the outcome of a public interest balancing process in which the benefits of the project are balanced against the detriments. A permit is granted unless the proposal is found to be contrary to the public interest.

The Corps periodically issues policy and guidance that govern its administrative procedures for the granting of permits, public review, interagency review, and enforcement of the Section 10 and Section 404 regulatory programs. After a permit is issued, the Corps may visit the construction site to determine compliance with the permit decision. If a Section 404 permit requires compensatory mitigation, the permit applicant may be instructed to conduct periodic environmental monitoring. If the Corps discovers that permit conditions were violated by the permittee, it may issue a compliance order, suspend or revoke the permit, or initiate civil judicial action against the permittee (NRC, 2001a). The information re-

TABLE 3-1 U.S. Army Corps of Engineer's Regulatory Program, Total Permit Decisions, FY 2002

Permit Decisions	Number
Standard permits[a]	4,023
Nationwide[b]	35,768
Regional[c]	38,125
Letter of permission[d]	3,258
Denial[e]	128
Withdrawn[f]	4,143
TOTAL	85,445

[a] Permits that require public notice, opportunity for public hearing, and an analysis of project alternatives and completion of an environmental assessment.
[b] General Permits issued by Corps Headquarters to authorize activities with minimal impacts across the country.
[c] General Permits issued by division or district engineers to authorize activities in particular geographic areas.
[d] Permits issued where it is determined by the district engineer that the proposed work would be minor and have no significant impact on the environment.
[e] Applications denied with or without prejudice:

- Denial with prejudice occurs when a permit is denied because it is contrary to the public interest or result in unacceptable environmental impacts.
- Denial without prejudice occurs when a permit is denied because the applicant lacks the necessary approval, such as water quality certification.

[f] Individual permit applications withdrawn by the applicant or by the Corps.

SOURCE: U.S. Army Corps of Engineers (2003c; available [on-line] at *http://www.usace.army.mil/inet/functions/cw/cecwo//reg//2002web charts.pdf* [accessed March 24, 2004]).

presented by these permits forms an extensive catalogue of activities within the nation's river basins and coastal systems.

THE U.S. ARMY CORPS OF ENGINEERS ROLE IN ENVIRONMENTAL STEWARDSHIP

The Corps controls or strongly influences an enormous range of activities that alter ecological systems, watersheds, and riverine and coastal systems, both through the projects that the Corps plans and implements and through the permits it issues. The potential of the Corps to alter the structure and functioning of the nation's ecosystems is immense. The Corps, itself, acknowledges that it must "move away from and avoid a way of doing business that would contribute to greater irreversible changes to natural ecosystems, and instead, move us toward environmental and ultimately, economic and social improvements" (U.S. Army Corps of Engineers, 2002b, p. 3).

Federal agencies are being challenged to look beyond the immediate impact of their activities and promote land and water management actions that increase overall ecosystem health. For example, the Corps, which is part of the Department of the Army, was one of 14 signatory agencies of a 1995 memorandum of understanding to foster the ecosystem approach. The stated goal of the ecosystem approach is "to restore and sustain the health, productivity, and biological diversity of ecosystems and the overall quality of life through a natural resource management approach that is fully integrated with social and economic goals" (U.S. Army Corps of Engineers, 1999a, p. 16). Thus, the Corps, along with other agencies, is being charged to go beyond minimizing harm to enhancing environmental quality and living resources. Because of the Corps' activities, expertise, authorities, and infrastructure, its staff is in a unique position to enhance environmental quality and should seek opportunities for environmental stewardship as a part of all projects and permitting activities—not only when it is requested to do so by state and local partners. This emphasis on environmental stewardship should not be limited to restoration projects, but instead should be a major consideration in planning, design, and decision making for all Corps activities.

The federal *Economic and Environmental Principles for Water and Related Land Resources Implementation Studies* (*P&G*) (WRC, 1983), signed by President Reagan in 1983, guides Corps' pre- and post-authorization studies and requires that alternative plans be "formulated in consideration of four criteria . . . completeness, effectiveness, efficiency, and acceptability" (U.S. Army Corps of Engineers, 2000c, p. 2-4). Given the emphasis the Corps and other agencies are placing on environmental stewardship, the following are suggestions by members of the panel on how to retain the important role of environmental stewardship in project planning, construction, and perhaps permitting. This will require, but should not be limited to, the following:

- Consideration and evaluation of opportunities for ***environmental stewardship***, including restoration, ***in all Corps projects***

- Consistent consideration of ***the cumulative effects of all human activities***, in water project planning in river basins and coastal systems; such consideration should take into account not only the spatial scales over which cumulative impacts may occur, but also the time frame over which these impacts might occur; guidelines to determine the spatial and temporal framework required in project evaluation should be incorporated into the *P&G* or documented in Corps policies and procedures
- ***Expansion of the spatial and temporal scales*** of the system to be analyzed in connection with permitting and projects, and the application of a ***watershed approach*** to planning (thereby allowing full consideration of the geographic scope of benefits, the environmental effects of planned projects and permits, and the complementary and antagonistic effects of multiple activities within a watershed)
- ***Improvement of evaluations to measure the effectiveness of mitigation and restoration efforts and the adoption of an adaptive management approach to restoration*** including increasing Corps expertise in ecosystem restoration
- ***Improvement of interagency collaboration and coordination*** on actions affecting natural resources (see Chapter 5)
- ***Improvement of methods for balancing environmental and economic costs and benefits*** in project planning and evaluation (see Chapter 4)
- ***Protection of sensitive and ecologically important habitat, species, and ecosystem functions***

Many of the above issues and challenges to fully integrating a sound program of environmental stewardship into the Corps' activities have been acknowledged and acted on by the Corps. Announcement of the new Corps Environmental Operating Principles (Box 3-3) by Chief of Engineers Lt. General Robert Flowers marked an important step in increasing the role of environmental stewardship in Corps activities.

The challenge that the Corps must now address is the thorough integration of these principles into its operating and evaluation procedures, and the development of mechanisms to foster a strong, uniform emphasis on environmental stewardship throughout the agency. A protocol for prioritizing projects for implementation could be developed and could take into account not only social needs and costs, but also environmental effects (positive and negative) and the effects of implementing or not implementing other projects within the watershed.

The *Program Management Plan for Integrating the Environmental Operating Principles Within HQUSACE* describes an extensive list of ongoing and planned activities to increase the focus on environmental stewardship within the Corps (U.S. Army Corps of Engineers, 2003c). Several of the issues highlighted in the plan have also been identified in this report, including: improved training and education on the concept of sustainability; improved planning methods to achieve a greater balance between economic and ecosystem benefits attributed to projects; and increased interagency cooperation and collaboration with nongovernmental organizations with a specific focus on environmental protection and restoration. Based on the arguments and logic presented in this chapter and in this report as a whole, the Corps is encouraged to progress as rapidly as possible in meeting its goal of incorporating "environmental restoration and sustainability" into the planning and implementation of all Corps projects (U.S. Army Corps of Engineers, 2003c, p. 22).

The September 2002 Corps of Engineers strategic plan expresses the Corps' commitment to fully integrate environmental stewardship and restoration into its activities (U.S. Army Corps of Engineers, 2002c). As part of its commitment to a sustainable future, the Corps acknowledges shifting national values assert that "growth and development must occur in a sustainable manner so as to protect vital ecosystems" and it accepts "responsibility for the condition of the environment and natural resources through . . . stewardship, regulatory, project planning, engineering, construction, and operations activities" (U.S. Army Corps of Engineers, 2002c, p. 6). Two of the five strategic goals are (1) providing sustainable (including environmentally sustainable) development and integrated management of the nation's water resources, and (2) repairing past environmental degradation and preventing future environmental losses—and the Corps' associated objectives and initiatives offer the potential to position it as a key player in the protection and restoration of our nation's environment.

The environmental stewardship record of the Corps is mixed and includes both successes and failures. Some of the most ambitious and costly restoration projects (e.g., the Kissimmee River, the Everglades, and the Louisiana coast) are designed to correct damage caused by previous Corps' projects that were conducted at a time when controlling water regimes was far more important to society than environmental stewardship. The challenge to the Corps will be to make substantive changes in procedures that have not historically made a positive contribution to stew-

BOX 3-3
Environmental Operating Principles of the U.S. Army Corps of Engineers

1. Strive to achieve environmental sustainability. An environment maintained in a healthy, diverse, and sustainable condition is necessary to support life.

2. Recognize the interdependence of life and the physical environment. Proactively consider environmental consequences of Corps programs and act accordingly in all appropriate circumstances.

3. Seek balance and synergy among human development activities and natural systems by designing economic and environmental solutions that support and reinforce one another.

4. Continue to accept corporate responsibility and accountability under the law for activities and decisions under Corps control that impact human health and welfare and the continued viability of natural systems.

5. Seek ways and means to assess and mitigate cumulative impacts to the environment; bring systems approaches to the full life cycle of Corps processes and work.

6. Build and share an integrated scientific, economic, and social knowledge base that supports a greater understanding of the environment and impacts of Corps work.

7. Respect the views of individuals and groups interested in Corps activities, listen to them actively, and learn from their perspective in the search to find innovative win-win solutions to the nation's problems that also protect and enhance the environment.

SOURCE: U.S. Army Corps of Engineers (2001a; available [online] at http://www.hq.usace.army.mil/cepa/envprinciples.htm [accessed March 25, 2004]).

ardship of our nation's environment. This will require support from Congress as well as leadership and support from within the Corps. Chapter 5 considers various barriers and possible remedies to more fully and consistently implementing an integrated approach to water resources management—one that incorporates environmental stewardship.

SUMMARY

The extended time period from the request by a sponsor to project execution and post-project evaluation, combined with the rapidly changing importance of environmental restoration and stewardship in the Corps' agenda, make it difficult to separate historical problems from current procedures. Some (mostly older) Corps projects have caused great and, in some cases irreparable, harm. Other (mostly newer) Corps projects can be seen as demonstrable successes in promoting environmental restoration. The potential for increasing success of the Corps as an important agent for environmental stewardship and restoration is illustrated in the following documents: the Environmental Operating Principles (U.S. Army Corps of Engineers, 2001a); the *Program Management Plan for Integrating the Environmental Operating Principles Within the HQUACE* (U.S. Army Corps of Engineers, 2003c); the Corps Environmental Operating Principles and Implementation Guidance (U.S. Army Corps of Engineers, 2002g); and the draft strategic plan (U.S. Army Corps of Engineers. 2002c).

The Corps and other federal agencies have been charged with fostering an "ecosystem approach," which seeks to integrate social and economic goals with the restoration and preservation of natural ecosystems. Simply minimizing harm to the environment is, therefore, no longer sufficient. The Corps should endeavor to improve environmental quality in all of its projects (not just its restoration projects). Potential impacts should take into account not only the spatial scales over which cumulative impacts may occur, but also the time frame over which they might occur.

Recommendation 3-1: The Corps should ensure that all project plans include an assessment of how the project fulfills the Corps' commitment to environmental stewardship. The cumulative effects of each project, together with other past and future human activities in the same river basin or coastal system, should be consistently evaluated for all projects.

Stewardship of hydrologic systems and the ecosystems they support requires that water resource projects be designed, implemented, and evaluated in a way that accounts for economic as well as environmental objectives at appropriate temporal and spatial scales. The analysis necessary to identify and appropriately evaluate these objectives and trade-offs forms the foundation of integrated water system planning. Chapter 4 examines the basis and methods of Corps integrated water project planning within the river basin and coastal system context.

4

Authorities, Methods, and Practices of Integrated Water Project Planning in River Basins and Coastal Systems

Chapter Highlights

This chapter describes the Corps' current approach to integrated water project planning and management within a river basin and coastal system framework, including its governing authorities, methods, and practices. A range of examples is provided in which integrated system planning was (or was not) successfully achieved.

THE CORPS' CURRENT AUTHORITIES AND PROCEDURES

Authorities

Integrated river basin and coastal systems planning by the Corps is supported by a number of legislative authorities (Table 4-1). In addition, there are many specific study resolutions and authorities that permit, and in some cases specify, comprehensive examinations of water resources needs and opportunities at the river basin and coastal system scale.

Integrated water project planning is also indicated in a number of authorities supporting an ecosystem approach in Corps activities, whether

TABLE 4-1 Legislative Authorities Supporting Integrated River Basin and Coastal Systems Planning

Legislation	Purpose
River and Harbor and Flood Control Act of 1970, Section 216 (P.L. 91-611)	Authorizes investigations for modifying existing projects and their operations when found "advisable due to significantly changed physical or economic conditions and for improving the quality of the environment in the overall public interest"
Water Resources Development Act (WRDA) of 1974, Section 22 (as amended)	Authorizes cooperation with states and Indian tribes in preparing plans for the development, utilization, and conservation of water and related land resources of drainage basins, ecosystems, and watersheds
WRDA of 1986, Section 1135 (as amended)	Authorizes modifications in the structures or operations of water projects for the purpose of improving the quality of the environment in the public interest; this authority is limited to existing Corps projects
WRDA of 1992, Section 204	Authorizes the beneficial use of dredged material for the protection, restoration, and creation of aquatic and ecologically related habitats
WRDA of 1996, Section 206	Authorizes aquatic ecosystem restoration and protection projects to be pursued at sites with no existing Corps project
WRDA of 1996, Section 207	Authorizes the selection of a dredge material disposal method that is not the least-cost alternative in order to achieve environmental benefits, to include creation of wetlands and shoreline erosion control for the purpose of protecting significant ecological resources

TABLE 4-1 Continued

Legislation	Purpose
WRDA of 1996, Section 221	Added watersheds and ecosystems, providing the opportunity for Section 22 authority (WRDA of 1974) to be used for watershed and ecosystem studies
WRDA of 2000, Section 202	Provides authority for the Corps to assess the water resource needs of river basins and watersheds, including ecosystem protection and restoration, flood damage reduction, navigation and ports, watershed protection, water supply, and drought preparedness

ecosystem restoration is the sole objective of the project or one of many. These authorities define compliance requirements, emphasize protection of environmental quality, and endorse federal efforts to advance environmental goals. They include specific authorizations for individual project reconnaissance and feasibility planning as well as programmatic authorities that allow projects for the purposes of restoring and protecting ecological resources to be carried out without specific congressional authorization (e.g., Section 1135 of the Water Resources Development Act [WRDA] of 1986, as amended [P.L. 99-6662] authorizing "Project Modifications for Improvement of Environment"; and Section 206 of the WRDA 1996 authorizing "Aquatic Ecosystem Restoration" [P.L. 104-303]). Unlike Section 1135 projects, whose activities are limited to sites with existing Corps projects, Section 206 projects may be pursued at sites with no existing project. In addition, two WRDA sections authorize beneficial use of dredged material—Section 204 of the WRDA of 1992, as amended (P.L. 102-580), and Section 207 of the WRDA 1996 (P.L. 104-303; Table 4-1).

Section 216 of the River and Harbor and Flood Control Act (P.L. 91-611) authorizes investigations for modifying existing projects and their operations when found "advisable due to significantly changed physical or economic conditions and for improving the quality of the environment in the overall public interest." After an initial appraisal, the so-called 216 study process can lead to a general investigations study, which can be appropriate for large-scale ecosystem restoration projects linked to existing civil works projects, but whose costs would be too high for Section 1135, Section 206, or Section 204 authorities.

In 2000, the Corps joined other federal land and water management agencies in adopting a unified federal policy on watershed management (60 Fed. Reg. 62566). The policy has two stated goals: (1) use a watershed approach to prevent and reduce pollution of surface and groundwaters resulting from federal land and resource management activities; and (2) accomplish this in a unified and cost-effective manner. This policy follows a 1995 memorandum of understanding (MOU) among federal land and water management agencies signed to foster an ecosystem approach to natural resource management, protection, and assistance (Appendix A in Corps Engineering Pamphlet No. EP 1165-2-502; available [on-line] at *http://www.usace.army.mil/inet/usace-docs/eng-pamphlets/ep1165-2-502/* [accessed March 24, 2004]). This 1995 MOU defines an ecosystem approach as "a method for sustaining or restoring ecological systems and their functions and values." It specifically integrates ecological, economic, and social factors and states that these are to be "applied within a geographic framework defined primarily by ecological boundaries." The 1995 MOU states that federal agencies should ensure the utilization of authorities in a way that facilitates, and does not pose barriers to, an ecosystem approach.

U.S. Army Corps of Engineers Planning Guidance

An emphasis on integrated watershed and coastal systems planning is also found in recent Corps policy guidance. The *Planning and Guidance Notebook* (*PGN*) states that "civil works planning should incorporate a watershed perspective, whether that planning involves a project feasibility study or a more comprehensive watershed study" (U.S. Army Corps of Engineers, 2000b, Section 2-6) and "every effort shall be made to assure that both economic and environmental value is added to watershed resources" (U.S. Army Corps of Engineers, 2000b, Section 2-1). The *PGN* also states that planning "should consider the sustainability of future watershed resources, specifically taking into account environmental quality, economic development and social well-being" (U.S. Army Corps of Engineers, 2000b, Section 2-6). Guidance on environmental restoration and protection is summarized in the Digest of Water Resources Policies and Authorities (U.S. Army Corps of Engineers, 1999b) and Ecosystem Restoration: Supporting Policy Information (U.S. Army Corps of Engineers, 1999a). Strong support for watershed and coastal system planning and ecosystem restoration and protection has also been expressed recently in public testimony by both the Chief of Engineers Lieutenant General R.B Flowers and the Director of Civil Works Major General Robert Griffin (Griffin, 2002; U.S. Senate, Committee on Environment and Public Works, 2002).

Despite the clear authority and strong evidence for internal support for integrated river basin, coastal system, and ecosystem planning, the amount of focused guidance for such planning is relatively limited compared to that for economic evaluation of water projects. Basic requirements for planning Corps projects were formalized by Congress in the WRDA 1986 (P.L. 99-662), which requires two levels of study for each Corps project: a reconnaissance study and a feasibility study (Box 4-1). These two levels of study are reflected in the *PGN*, which remains the primary planning guidance for all Corps projects. The current version of the *PGN* (dated April 22, 2000) contains six steps that must be followed by the Corps when planning water resources projects (U.S. Army Corps of Engineers, 2000c, p. 2-3):

Step 1: Identify problems and opportunities
Step 2: Inventory and forecast conditions
Step 3: Formulate alternative plans
Step 4: Evaluate alternative plans
Step 5: Compare alternative plans
Step 6: Select a plan

Step 1 of the planning process is to identify the problems and opportunities that reflect the priorities and preferences of the local sponsor, the federal government, and others participating in the planning study. This "scoping" process leads to a statement of the "planning objectives" that describe the desired outcome of the plan. Step 2 is to develop an inventory and forecast relevant resources in the planning area under current and future "without-project" conditions.

The first phase of the plan formulation process (Step 3) is identification of the structural and nonstructural management alternatives that would achieve the objectives of study participants. Management alternatives are then evaluated (Step 4) by forecasting the most likely "with-project condition" expected under each alternative plan (U.S. Army Corps of Engineers, 2000c, p. 2-6). This includes a characterization of the beneficial and adverse social, environmental, and economic effects of each alternative. In Step 5, the Corps compares the beneficial and adverse effects of the various alternative plans (including the no-action alternative). After making this comparison, the plans are ranked. The process culminates (Step 6) in the selection of a project plan from the alternatives or the decision to recommend that no action be taken. The criteria for selecting the recommended plan depend on the type of plan and whether the project outputs support national economic development (NED), national ecosystem restoration (NER), or a combination of both. With the exception of projects explicitly focused on ecosystem restoration (therefore supporting NER), Corps policy mandates the most cost-effective (highest net benefits) implementable alternative as the alternative, described in Box 4-2 (U.S. Army Corps of Engineers, 2000c, p. 2-10).

Neither spatial integration of projects within a watershed nor evaluation of the cumulative effects of multiple projects within a watershed is incorporated as an explicit element of the six project planning steps in the *PGN*. The project planning process is limited to the planning area as defined in the federal *Economic and Environmental Principles and Guidelines for Water and Related Land Resources Implementation Studies (P&G)*: "the planning area is a geographic space with an identified boundary that includes: the area identified in the study's authorizing document" (WRC, 1983, Section 1.4[7]). Hence, an evaluation of projects in the upper reaches of a river basin is not explicitly required to consider downstream effects or the cumulative impact of multiple, independent projects. Requirements in the *P&G* for mitigating adverse effects of projects and the "scoping process" under the National Environmental Policy Act (NEPA) (40 Fed. Reg. 1500-1508) do not explicitly address system-wide effects of projects within a watershed or coastal system. Alternative project plans can be compared based on their beneficial and adverse effects within the immediate project domain. In the sections of the *PGN* that provide supporting guidance for each of the six project planning steps, there is a single paragraph (U.S. Army Corps of Engineers, 2000b, Section 2-6) that calls for "a watershed perspective." Unlike the detailed instructions for cost-benefit analysis, public participation, and cost-sharing, this paragraph contains four general statements about the need to take into account "the interconnectedness of water and land resources" (U.S. Army Corps of Engineers, 2000b, Section 2-6). Procedures for implementing a watershed or coastal system perspective are not identified, nor are any other documents referenced for this purpose in the *PGN*. Moreover, in some cases, the existing guidance may, albeit unintentionally, conflict with the objective of taking a broader systems approach, as in the case of the language that effectively encourages the comparison of alternative project plans based on their beneficial and adverse effects within the *immediate* (e.g., local) project domain.

BOX 4-1
The Corps' Project Cycle

After a potential problem has been identified either by the Corps or by a local authority, the Corps must first determine the extent of the problem and the likelihood that it can implement a solution. This is called the reconnaissance phase; and it is 100 percent funded by the federal government, must be completed within one year, and cannot exceed $100,000. The Corps district office performs this study, which also includes identifying a prospective local sponsor for cost-sharing during the feasibility phase. The reconnaissance study will recommend either that a feasibility study be conducted or that all further planning be discontinued. If the decision is made to continue on to the feasibility phase, the Corps and the local sponsor will negotiate a project study plan (PSP) and a feasibility cost-sharing arrangement (FCSA). The PSP will include the scientific, engineering, and management activities consisting of a detailed agreement of the task descriptions, task responsibilities, and task milestones.

The cost-sharing responsibility of the local sponsor is nonnegotiable as outlined in the WRDA 1986 (P.L. 99-662). The federal government will split the cost of the activities of the feasibility phase 50-50 with the local sponsor, who may substitute 25 percent of its 50 percent share with in-kind products and services. The intent of the in-kind contribution was to assist communities with financial difficulties in meeting their 50 percent cost-share obligation. Many communities make use of this in-kind opportunity.

Once it is determined that the Corps' service is needed and a local cost-sharing sponsor has signed the FCSA, the Corps will hold at least one public workshop. Thus begins the feasibility phase, which has no time restriction but typically lasts approximately four to five years. At this stage it is determined whether a federal project is appropriate for solving the identified water resource problem(s). During this phase the Corps receives public input, and alternative plans are developed and revised. Project design characteristics are outlined, project costs are estimated, benefit-cost ratios are determined, and the national economic development (NED) alternative is identified (see Box 4-2 for explanation of the NED alternative). In some cases the local sponsor

continued

BOX 4-1 Continued

may decide on an alternative that exceeds the NED alternative (e.g., desire for protection against a 200-year flood level, rather than a 100-year level), in which case the local sponsor must agree to cover the additional costs if the Corps agrees to proceed with the local sponsor's suggested alternative (U.S. Army Corps of Engineers, 2001b).

After the alternative plans have been developed, the Corps may convene an alternative formulation briefing (AFB) to present the plan alternatives, the NED plan, and the recommended alternative plan. After the AFB, the Corps will complete its draft feasibility report and submit it to Corps Headquarters and the Secretary of the Army, as well as other relevant federal agencies and the U.S. Office of Management and Budget. The public will then have 45 days to review the report and the Corps will hold a final public meeting to obtain further comments. Within approximately six months of submitting the AFB, the Corps will release a revised feasibility report. During this time the Corps is also concluding the National Environmental Policy Act's required environmental impact statement (40 Fed. Reg. 1500-1508). The feasibility phase ends when the local sponsor and the Corps decide on a final plan and sign the division engineer's notice, and this notice recommends to the Corps' chief of engineers that the plan should be approved.

Directly after signing the division engineer's notice, the district engineer will complete a preconstruction engineering and design (PED) cost-sharing agreement. This outlines the required preliminary specifications and identification of the lands, easements, rights of way, relocations, and disposal areas (LERRDs). This process takes about two years to complete. The PED is then sent to the Secretary of the Army, and the Corps' chief of engineers recommends approval of the project.

An important step would therefore be to review these main planning guidance documents and identify cases in which better and more explicit language would bring guidance for day-to-day project planning into closer conformity with general statements of principle and would clearly endorse planning on broader spatial and temporal scales. This effort should draw on the practical insights available from previously implemented Corps projects, where explicit planning on the scale of watersheds and/or coastal systems was carried out.

BOX 4-2
NED Benefits and the NED Account

The concept of "NED benefits" and the incorporation of these benefits into a "NED account" currently provide the foundation for the evaluation of proposed water resource projects undertaken by the Corps.

The procedures, principles, and practices for evaluating Corps water resource projects come from two main sources. One is the federal *Economic and Environmental Principles and Guidelines for Water and Related Land Resources Implementation Studies*, also known as the *P&G*, published in 1983, which provides a set of general principles and practices to be followed by the Corps as well as by other federal water-planning agencies (WRC, 1983). The second is the *Planning Guidance Notebook* (*PGN*), a Corps document (U.S. Army Corps of Engineers, 2000b). The *PGN* requires the Corps to follow a planning process that includes formulation, evaluation, comparison, and selection of alternative plans for water resource projects. In all cases, alternative plans must be evaluated and compared with reference to the federal objective. This objective is to "contribute to national economic development (NED) consistent with protecting the nation's environment, in accordance with national environmental statutes, applicable executive orders, and other Federal planning requirements" (U.S. Army Corps of Engineers, 2000b, Section 2-2 [a]). The results of this analysis are presented in terms of the NED account for the project. Benefits in the NED account are intended to show how the plan increases the production and/or consumption potential of the national economy, while costs in the NED account are intended to measure the national or social opportunity costs of implementing the plan.

For all project purposes except ecosystem restoration, "the alternative plan that reasonably maximizes net economic benefits consistent with protecting the nation's environment, the NED plan, is to be selected" (U.S. Army Corps of Engineers, 2000b, Section 2-2, f [1]). In practice, however, exceptions to this selection criterion are permitted if "the Secretary of a department or head of an independent agency grants an exception when there is some overriding reason for selecting another plan, based upon Federal, state, local, and international concerns" (U.S. Army Corps of Engineers, 2000b, Section 2-2, f [1]).

continued

BOX 4-2 Continued

The guidance for measuring NED benefits and costs provided in the *P&G* and the *PGN* is based on standard principles of benefit-cost analysis for measuring social benefits and costs of public programs. As a result, the "N" in NED signifies that benefits and costs outside the "private or commercial" realm are to be counted, not just those benefits or costs experienced by private parties, such as the cost of purchasing land. For example, when land is donated for use in a water project, the *P&G* requires that a NED cost be estimated because the land has value for alternative uses, even though the land has no budgetary cost. Similarly, the *P&G* requires that NED benefits include "the value of output resulting from external economies caused by a plan."

As noted in a recent National Research Council (NRC, 1999b) report, the basic economic guidance provided in the *P&G* for identifying and estimating NED benefits and costs is "largely sound." However, the same report also notes that there is a need for broadening and updating the *P&G*.

An area of particular concern is Corps practice for incorporating a range of environmental effects in water project planning. On one hand, both the *P&G* and the *PGN* recognize that water projects may have a range of environmental effects, and the guidance provided by the *PGN* makes it clear that such effects are to be an explicit part of the overall project planning process. On the other hand, there remains disagreement about whether and how to attach monetary measures to these environmental effects. Currently these effects are not included as project benefits and costs to be evaluated along with more traditional NED benefits and costs. Nonetheless, environmental effects can indirectly affect the NED assessment of a project when environmental considerations cause constraints on project design, which in turn can affect NED project benefits or costs.

In some instances, drafting more explicit guidance on how to incorporate broader systems effects into project planning involves nothing more than bringing practice into line with the traditional policy objective of maximizing the NED benefits from water projects. For example, the fact that evaluation of projects in the upper reaches of a river basin is not explicitly required to consider downstream or cumulative effects does not mean that such effects might not be considered. At the same time, the absence of specific instructions can make it more difficult for a project manager to justify the effort involved, while also leaving open the option of not considering these costs. As a result, the evaluation can lead to incorrect estimates of the true NED benefits and costs of upstream water projects from a social benefit-cost standpoint as illustrated in Box 4-3.

Recommendation 4-1: The Corps should undertake a thorough review of the documents issued to guide water resource planning "in the field" to bring such guidance into closer conformity with generally stated principles that support integrated water resources planning.

When performing this review, it is important to keep in mind that no two watersheds or coastal systems are alike in their hydrologic, ecologic, social, and economic conditions; therefore, creating specific instructions for integrated systems planning is neither advisable nor achievable. Nonetheless, clear guidelines regarding identification of the suite of potential objectives and impacts of a water project, and the basis for weighing the costs and benefits and developing effective trade-offs, are necessary for effective integrated water resources planning.

Practice

Although the Corps' planning documents lack comprehensive guidance on performing integrated river basin and coastal system studies, in practice the Corps is often able to conduct such analyses. District offices can call on general water resource planning principles, guiding authorities, and regulations to support development of an integrated plan. Doing so requires some initiative on the part of the individual project managers not only to develop an approach, but also to establish the necessary authority and, typically, to generate supplemental funding. Individual initiative may also be called upon because such efforts may be performed at the expense of other planning activities, activities that may lead more directly to funded construction projects and to addressing sponsor demands.

Typically, integrated water planning investigations at a watershed or coastal system scale are mandated by specific authorizing legislation. When given such authorization, the Corps has proven that it is capable of identifying salient issues, incorporating stakeholder interests, and building sophisticated models representing system-wide water flow and sediment loads, as well as socioeconomic aspects of proposed projects. Examples of these capabilities are given later in this chapter.

BOX 4-3
NED Benefits of Adopting a Comprehensive Framework

River basins and coastal waters are economically as well as physically and biologically interconnected. For example, flood control projects constructed upstream may affect the likelihood of flooding downstream and influence the need for dredging and shore protection in coastal streams. A series of local water projects undertaken separately for local sponsors may have cumulative effects on other activities, such as fisheries and agriculture elsewhere in the watershed.

Interactions of the sort described above affect the NED benefits and costs of publicly financed water projects. Consider the hypothetical case of a flood control project proposed at site A in a watershed. Assume that two alternative flood control projects are under consideration and that the estimated local NED benefits and NED costs of each alternative are shown in the table below. If the choice of alternative is made only with reference to estimated local NED benefits and costs, then alternative 1 would be selected as the "alternative plan that reasonably maximizes net economic benefits"(U.S. Army Corps of Engineers, 2000b, Section 2-3).

Suppose, however, that implementing alternative 1 would have negative economic effects in another part of the watershed not included in the local plan. In principle these costs should also be included as NED costs because they reflect "the value of that which is foregone when a choice of a particular plan or measure is made" (U.S. Army Corps of Engineers, 2000b, Section 2-4). Doing so would cause Alternative 2, which does not entail these costs, to be identified as the NED alternative plan.

Comparison of NED Benefits and Costs (millions of dollars)

	Local NED Benefit	Local NED Cost	Local Net NED Benefit	Watershed Effects	Watershed Net NED Benefit
Alternative 1	100	80	20	-20	0
Alternative 2	90	80	10	-0	10

continued

BOX 4-3 Continued

On one hand, sections of the *PGN* recognize that costs incurred outside the boundaries of the immediate local project should be taken into account. "Negative externalities" are listed as a component of "other direct costs" that should be included in estimated NED economic costs (U.S. Army Corps of Engineers, 2000b, Section 2-4), and the section of the *PGN* dealing with flood control projects states that "[d]ownstream consequences of dams on flood risk are also analyzed in a risk-based framework" (U.S. Army Corps of Engineers, 2000b, Section 3-3).

On the other hand, failure to explicitly analyze the effects of individual projects as part of a broader system can increase the chance that potentially important external costs (or benefits) of specific projects may be overlooked. In such cases, incomplete accounting of the full NED benefits and costs can prevent projects with the greatest *watershed* net economic benefits from being selected.

If project authorization and associated funding directives do not mandate examining the broader impacts of a project, including the cumulative effects of related projects, pressures to efficiently complete a study and satisfy sponsor expectations can narrow project focus to local costs and benefits, leading to cursory river basin or coastal system planning. Investigation may focus on immediate with- and without-project conditions to evaluate risks, costs, and benefits of a project. Broader investigations and modeling (such as those that integrate water quality, changes in rainfall patterns and sea level, and habitat response and ecosystem functioning, as well as the cumulative effects of other projects and permits) within a watershed generally require specific effort and costs and introduce possible project delays. Further, development and verification of such models are not generally the purview of Corps project managers and district engineers. Therefore, an outstanding challenge for the Corps is to incorporate integrated, systems-level analysis in the evaluation of routine water projects.

Spatial integration of water resources projects among Corps districts at the division level is informally addressed by a Policy Review Board. The Policy Review Board (consisting of district office representatives and the division-level executives) is given the opportunity to review projects within a watershed, but the review is informally structured to raise issues to review plans for projects within the division.

Within Corps divisions and districts, project managers are expected to provide the coordination necessary to ensure that projects will not adversely affect one another. Turnover in personnel can work against consistent integration among projects. Procedures for project integration (with the exception of the use of hydraulics models) rely chiefly on communication among project engineers and planners. Managers of three different branches of the New Orleans District's Division of Planning, Programs and Project Management indicated the need for geographic information systems (GIS) and habitat models to simulate project effects on water quality, storm surge, habitat variables, water level, and stream flow. Predictive models and tools of this nature are not currently prescribed in the federal *P&G* or other more specific guidance documents.

Several procedures have been adopted by the Corps to enhance project planning and, indirectly, multi-project integration. In 1993, the Office of Federal Procurement promulgated rules for a procedure called "value engineering" (Office of Management and Budget [OMB] Circular No. A-131, May 21, 1993). These rules, authorized by the WRDA of 1986 (P.L. 99-662) require that federal departments and agencies use value engineering as a management tool to reduce program and acquisition costs. In response, the Corps promulgated procedures to provide oversight during the planning of all projects with estimated costs of $1 million or more. The Corps' value engineering policy requires an independent assessment of each major project during both the planning and the design phases. The "value engineer" is a Corps employee not assigned to the project being evaluated. The value engineering team often includes employees from other Corps district offices. Another procedure that may enhance project planning and, indirectly, multi-project integration is the project management business process, which is based on a stated Corps commitment to provide "customer service" in delivering quality projects and includes elements of coordinated management, teamwork, partnering, and balancing of competing demands. Although not explicitly designed to foster integrated river basin and coastal system project planning, the guidelines for both of these programs could be revised to include system integration and environmental stewardship as explicit review objectives. These programs also provide useful experience for the Corps in evaluating operational models for program integration.

For each major Corps project, an independent technical review (ITR) team or committee is also appointed. The ITR is composed of a group of independent technical professionals (external to the project) who examine project designs that are complete or near complete to see if improvements or cost savings are possible. The main goal is to determine if the current design for the project satisfies the objectives and to see if there is a more effective or efficient way to accomplish the objective of the project. Like value engineering, the ITR is project specific and does not ensure integration or compatibility within a watershed. However, experience gained through both of these mechanisms may be useful in designing an approach for integrating Corps projects within river basins and coastal systems.

Watershed studies are authorized by the *PGN* (U.S. Army Corps of Engineers, 2000b, Section 3-9 [c]) and are defined as "planning initiatives that have a multi-purpose and multi-objective scope and that accommodate flexibility and collaboration in the formulation and evaluation process." These studies require a 50 percent cost-share from a local sponsor, which has limited the use of this provision for watershed-level planning. Similarly, multi-purpose flood mitigation and riverine restoration are described in the *PGN* (U.S. Army Corps of Engineers, 2000b, Section 3-9) and authorized by Congress in the WRDA 1999 (P.L. 106-53, Section 212). However, no funds were appropriated for this program in fiscal year (FY) 2001 and no funds were included in the administration's budget for FY 2002 (U.S. Department of the Army, 2002).

Trade-Off Analyses

Balancing the economic and the environmental benefits and costs of water projects involves both identifying and evaluating potential trade-offs between these two goals.

Although effective water resource decision making at all levels (e.g., local as well as systems) requires an improved ability to evaluate these trade-offs, the need is likely to increase as the Corps moves toward broader and more integrated approaches to planning. Although the *PGN* recognizes the need for accounting for the environmental effects of water projects, the challenge of placing monetary value on environmental costs and benefits of water projects has prevented the Corps from weighing them directly along with the traditional economic benefits identified for inclusion in the NED account. Instead, these effects are included in the overall project plan through NER accounts.

The broad issue is not whether but how to incorporate environmental benefits and costs into the Corps' planning process. The *PGN* identifies ecosystem restoration as one of the "primary missions" of the Corps' civil works program and defines NER outputs as increases or decreases in the net quantity or quality of desired ecosystem resources (U.S. Army Corps of Engineers, 2001b, Section 2-2[b]). In cases in which ecosystem restoration is a primary objective, the *PGN* goes on to require that ecosystem plans are to be "formulated and evaluated in terms of their net contribution to increases in ecosystem value (NER outputs), expressed in nonmonetary units" (U.S. Army Corps of Engineers, 2001b, Section 2-2[b]).

In principle, including estimates of monetary benefits and the costs of changes in the availability and quality of environmental goods caused by water projects is consistent with the social accounting framework of benefit-cost analysis that underlies the preparation of the NED accounts. In practice, estimating the value of environmental goods is challenging.

Unlike many of the effects traditionally captured in the NED accounting framework, outputs of ecological restoration projects tend to include intangible values such as endangered species protection, and aquatic ecosystem protection or restoration, as well as aesthetic values (NRC, 1999b). Economists have developed a range of methods for assigning monetary values to such intangible effects, and the Corps has devoted considerable intellectual energy and resources to examine how these methods might be used in Corps planning and evaluation activities (NRC, 1999b).

Implementing such methods would effectively incorporate a form of environmental stewardship in traditional Corps planning activities by broadening the historical objective of maximizing net national economic benefits to include a range of formally monetized environmental benefits and costs. An advantage of adopting such an approach is that the environmental effects of Corps activities could then be directly evaluated with more traditional NED benefits and costs. Corps projects that improved the quantity and/or quality of environmental goods would increase net economic benefits, while actions that had the opposite effect would reduce net benefits.

Although the concept of subjecting all project effects to a common metric such as money may be intellectually appealing, there are both intellectual and practical reasons for treading cautiously. First, setting aside the issue of how to monetize nonmonetary effects, there is considerable scientific uncertainty about how to value and model nonmonetary effects. Second, although economists generally agree that many environmental effects can be translated into monetary terms, there continues to be debate on how this can best be accomplished. Last, carrying out methods to value and measure environmental goods can be costly in terms of both time and money, which may limit the ability of the Corps to routinely apply these methods, as it does more traditional NED measures of benefit and cost (NRC, 1999b).

Because of these obstacles, the Corps has taken alternative approaches to integrate NER and NED considerations into project planning in ways that are systematic, yet stop short of monetizing environmental benefits. In some watershed plans, environmental restoration has been included as part of the total NED cost of each alternative, as in the case of alternatives considered under the Sacramento-San Joaquin Watershed plan.

A disadvantage of handling environmental effects in this manner is that otherwise superior alternatives can be overlooked by procedures that include the economic costs but not the economic benefits of environmental restoration. Beneficial use of dredged materials offers one example of how failure to include monetary benefits of environmental improvements can create less than desirable incentives. As illustrated above, one potential beneficial use of dredged materials is to create wetlands. However, current policies and practices do not allow dollar values to be assigned to the creation of such wetlands. Failure to assign monetary benefits to wetland creation as a "beneficial use" of dredged materials has two potential consequences. At the margin, this practice makes it somewhat harder to justify incurring the extra disposal costs associated with using dredged material for wetland creation *in financial terms* because there is no monetary benefit that can offset these costs. It also may provide financial incentives to favor uses with benefits that can be more readily monetized.

One potentially promising approach that has recently been used in some projects is the application of formal analytical frameworks to aid decision makers in identifying trade-offs that may arise between changes in environmental quantity or quality in the NER accounts that are not expressed in monetary terms and to project net benefit or cost in the NED account.

In the case of the Central and Southern Florida Indian River Lagoon South Project, environmental and economic benefits and costs have been evaluated through the use of cost-effectiveness and incremental cost analysis. Under this procedure, the ecosystem benefits of alternative water projects were measured in terms of physical dimensions, such as acres of wetlands or habitat units. Estimates are then made of the economic costs of achieving these physical changes in output. The objective was to determine the least-cost means of achieving each possible level of environmental benefit, measured in physical units. The resulting lists of "least-cost plans" were then compared to each other to identify those plans that produced more of the environmental benefits at the same cost or at a lesser cost than other alternatives.

These and similar approaches to planning, which fall broadly under the rubric of "trade-off analysis," do not seek to arrive at a monetary "bottom line" that implicitly weighs both the environmental and the economic effects of projects. Rather, the purpose of a trade-off analysis is to present information about multiple attributes of projects (e.g., environmental and economic effects) in a consistent fashion that allows decision makers to better understand the range and scope of trade-offs between the NED and NER effects of different projects. Presenting such trade-offs allows the range of choices among competing projects to be identified clearly and systematically, while at the same time acknowledging that actually weighing these options against others requires making judgments about different and competing project attributes and taking into consideration science, values, and stakeholder interests.

The use of trade-off analysis as an analytical framework for integrating NER and NED effects into project planning offers a feasible alternative for allowing the Corps to engage in systematic project planning, given the continuing controversy over the reliability and validity of existing approaches for monetizing environmental benefits and costs. The Corps has developed extensive guidelines for how to undertake such analysis (Institute for Water Resources, 2002). At the same time, the example of how environmental benefits of wetlands are (at least implicitly) treated in the analysis of beneficial use of dredged materials (discussed later in this chapter) points to the advantages of developing at least some methods for valuing the economic costs and benefits of environmental restoration.

At the institutional level, the Corps is aware of the conceptual issues and current best practices for systematically including the environmental effects of projects along with the more traditional NED benefits and costs. The challenge, as noted by another panel of this National Research Council (NRC, 2004b) study, is to determine how best to include these issues and practices in the documents that guide day-to-day planning. This is of particular importance during water resource planning of large, complex multi-component projects (such as the Everglades restoration project) or of individual projects that may have an impact on large, complex systems.

Recommendation 4-2: The Corps should continue its current efforts to facilitate the systematic integration of information provided in the NED and NER accounts through the application of trade-off analysis.

Recommendation 4-3: The Corps should continue to take steps, such as those outlined in *New Directions in Water Resources Planning for the U.S. Army Corps of Engineers* (NRC, 1999b) and *Analytical Methods*

***and Approaches for Water Resources Project Planning* (NRC, 2004b) to develop protocols and standards for incorporating environmental benefits and costs into project planning in a manner that is comparable to traditional NED benefits and costs.**

In addition, planning projects at a watershed or coastal system scale is more likely to require the consideration of nonstructural as well as structural approaches to deal with issues such as flood control and beach erosion. The federal *P&G* requires that nonstructural approaches to flood control be considered as alternatives to structural approaches. The *P&G* procedures for estimating the benefits and costs of nonstructural approaches, however, assume that differences in value between developed land and land left fallow as a nonstructural measure fully reflect the social value of developed and undeveloped land in floodplains, watersheds, or coastal areas. This assumption may not be correct and hence may unintentionally bias project choices in favor of structural approaches. Moreover, implementing nonstructural solutions in cases where development has already occurred can involve challenging issues raised by the need to compensate those who have made investments in floodplains under reasonable assumptions about government policy. The costs of such compensation may increase the cost of nonstructural projects without affecting their actual social cost since the required compensations are essentially redistributed from a society that gains from the nonstructural approach to those who suffer economic and social harm as a consequence.

EXAMPLES OF THE CORPS' USE OF INTEGRATED PLANNING IN RIVER BASINS AND COASTAL SYSTEMS

In reviewing Corps performance in using an integrated approach to water project planning in river basin and coastal systems, it is useful to examine instances in which the Corps has specifically completed such studies. These examples serve to illustrate the Corps' capabilities in this regard and, by extension, the conditions necessary to allow such studies to take place. Some examples also illustrate that integrated planning does not necessarily lead to integrated water resources management. Other factors, such as methodological, political, or jurisdictional conflicts can stand between the formulation and implementation of a coherent plan.

It is not difficult to find examples of projects in which the study objectives include a systems approach that specifically incorporates multiple scales, multiple objectives (including environmental stewardship), and diverse stakeholders. Many of the most noteworthy projects have a high profile, involve significant environmental and economic resources, and involve extensive field data sets, GIS support, simulation models, and decision support systems (Box 4-4).

With environmental restoration now established as a central Corps mission, smaller watershed studies are developed with the specific aim of identifying opportunities for that purpose. In collaboration with state and local jurisdictions, project authorities are developed with the objective of evaluating the condition of a watershed, often leading to a watershed plan that identifies local actions that may have beneficial regional effects. For example, in the Baltimore-Washington area, the Corps' Baltimore District is engaged in a variety of watershed studies motivated by regional objectives including reducing sediment and nutrient input to the Chesapeake Bay; restoring freshwater fisheries impacted by urbanization and by existing water projects for flood control, navigation, and water supply; and restoring aquatic and wildlife habitat. A watershed-based ecosystem restoration approach is used in these studies, leading to the identification of subbasin watersheds as potential project areas, rather than the traditional site-specific selections to identify potential restoration projects. Collaboration among the U.S. Environmental Protection Agency, the State of Maryland, the District of Columbia, the City of Baltimore, Maryland county governments, and other local authorities has resulted in additional broad objectives including water quality improvement, brownfield restoration, economic revitalization, and the public use and enjoyment of urban rivers. Interaction among diverse agencies can also work against development of specific projects. For example, despite agreement on broader objectives for ecosystem restoration, differences in specific priorities and preferred methodologies can halt a project, even after reconnaissance and feasibility studies identify worthwhile goals.

In the traditional water resource areas of flood damage reduction and navigation, integrated water resource planning must generally incorporate significant existing civil works. In these cases, integrated studies must typically balance the cumulative impacts of installed structures, the increased value now placed on ecosystem resources, and a constituency with a strong vested interest in maintaining the water resource benefits provided by installed works (an example of this issue is given in the case study that follows).

BOX 4-4
Examples of Corps Efforts at Integrated Water Resources Planning

Comprehensive Everglades Restoration Program
Louisiana Comprehensive Coastwide Ecosystem Restoration Study
Upper Mississippi Navigation Study
Upper Mississippi Comprehensive Plan
Missouri River Master Plan
Hudson River Study
Lower Snake River Salmon Passage Study
Sacramento-San Joaquin Rivers Comprehensive Plan
California Coastal Master Plan
Ohio River Main Stem Study
Great Lakes Study
Northern Gulf of Mexico Regional Sediment Management Demonstration Project

Case Study: Sacramento and San Joaquin River Comprehensive Plan. Following disastrous flooding on the Sacramento and San Joaquin Rivers in January 1997, California and federal legislation authorized the development of comprehensive plans for flood damage reduction and ecosystem restoration. Preparation of the plan was a joint effort of the Reclamation Board of California and the Corps. The goals of the Sacramento and San Joaquin River Comprehensive Study (comprehensive plan) were to reduce threats to public health and safety, reduce flood damages, and restore the ecosystem along the floodplain corridors. These goals expand on those of the original flood management system, developed over the twentieth century, which were to manage and redirect flood flows for economic (primarily agricultural) use and navigation. Water conveyance has superceded navigation as an important use, and ecosystem restoration has been added as an additional objective.

A major element of the comprehensive plan involves developing tools to describe the behavior of the complete flood conveyance system. A focus area is the modeling capability of demonstrating how changes in one part of the system will affect the performance of the system as a whole. The technical tools consist of computer models and an extensive information data base that support a system-wide approach.

The major findings of the plan are the following (California Reclamation Board and the U.S. Army Corps of Engineers, 2002):

- The system cannot safely convey the flows that it was formerly considered capable of accommodating.
- If levee reliability were improved system-wide, substantial increases in flood storage capacity would be necessary to avoid transferring increased flood risks to downstream areas.
- A comprehensive solution to improve public safety, reduce flood damages, and restore degraded ecosystems will require a combination of measures that increase conveyance capacity, increase flood storage, and improve floodplain management.

As originally conceived, the Sacramento-San Joaquin River Study lead to a single master plan specifying flood damage reduction and ecosystem restoration projects throughout California's Central Valley. Although the comprehensive plan reports wide agreement that the system requires improvement, it also concludes that there is little agreement on the measures that should be used and, therefore, does not recommend any particular projects. Instead, the comprehensive plan establishes a set of general principles to be used as a guide for future projects and establishes an approach to evaluate system-wide effects for all projects, regardless of scale. The plan also provides an administrative structure to oversee project analyses and to consistently apply emerging principles for maintaining the flood management system and developing future projects. The comprehensive plan also advocates a science-based adaptive assessment and management approach, with explicit incorporation of updated data bases, performance measures, coordination among resource management programs, peer review, and ongoing technical studies, including consideration of potential climate change. It also uses systems models as the basis for operation and maintenance of the existing systems as well as for future project development. As described in the interim report, the comprehensive plan establishes (1) a set of principles to guide future projects; (2) an approach to develop projects with consideration for system-wide effects; and (3) an organization to consistently apply the guiding principles in maintaining the flood management system and developing future projects.

Integrated Planning in the Mississippi River Basin

In the Mississippi River basin, integrated water planning becomes essential and faces the full range of jurisdictional hurdles. For more than 150 years, the Corps has developed and maintained an extensive system of levees, locks and dams, and channel modifications to support navigation and to control flood damage. Recent and ongoing examples of integrated planning efforts in this basin include the Upper Mississippi Navigation Study, the Upper Mississippi Comprehensive Plan, the Ohio River Main Stem Study, and ongoing attempts to revise the Missouri River Master Plan. As social values have shifted and recognition of the environmental degradation of the delta and the Gulf of Mexico has increased, the Corps' role has shifted from that of master engineer to that of manager within an enormously complex political and jurisdictional system. The case study below describes the evolution of the Corps strategy for ecosystem restoration within the Mississippi River Basin, and specifically within its lower portion—the Louisiana coastal zone. This case study shows that the Corps has become increasingly experienced in jointly planning environmental restoration projects, especially high-profile and complex restoration projects, with other agencies and stakeholders.

Case Study: Systems-Level Approach for Environmental Restoration in Coastal Louisiana. Many environmental problems in the Louisiana coastal area stem from human alteration of the geologic, hydrologic, and biologic processes that created the Mississippi River deltaic plain. Subsidence, sea-level rise, wetland loss, saltwater intrusion, landward migration of estuarine species, and barrier island erosion are all attributed to a combination of human-induced and natural processes. Louisiana has lost more than 1,500 square miles of coastal wetlands since 1956 (Figure 4-1). The conversion of wetlands to open water in coastal Louisiana accounts for about 80 percent of the coastal wetland loss in the continental United States.

The economic value of Louisiana's coastal wetlands varies depending upon what is included in the analysis, but estimates range from $2 billion to $17.9 billion per year. It is estimated that more than $100 billion in natural resources (including about one-third of the commercial fish and shellfish harvests in the contiguous United States and 20 percent of the nation's waterfowl) could be lost over the next 50 years if Louisiana's coastal wetlands continue to disappear at the present rate (Coalition to Restore Coastal Louisiana, 1999). These coastal wetlands are important natural defenses for the nation's energy infrastructure. Roughly 80 percent of the outer continental shelf's oil and gas is processed along the Louisiana coast, as well as two-thirds of the nation's imported oil. The wetlands also protect coastal communities (70 percent of the population of Louisiana) from storm surges. The annual storm protection value of Louisiana's coastal wetlands is estimated at between \$208 and \$904 per acre (Costanza et al., 1989).

Efforts to combat wetland loss in coastal Louisiana have been under way for roughly 35 years. In 1967, a U.S. House Resolution (H.R. 112, 86th Congress) directed the Corps (authorized under Section 3 of the Rivers and Harbors Act of 1902, 33 U.S.C. 418) to develop a coast-wide strategy for Louisiana "in the interest of hurricane protection, prevention of saltwater intrusion, preservation of fish and wildlife, prevention of erosion, and related water resource purposes." In subsequent years the Corps has planned and built freshwater diversions from the Mississippi River at Caernarvon and Davis Pond, with the State of Louisiana providing the local cost-share. In 1978, Louisiana established a state coastal zone management program that emphasized controlling activities that contributed to wetland loss. The state's program became a federally approved coastal zone management program in 1980. In 1981, the Louisiana legislature established the Coastal Environment Protection Trust Fund and appropriated \$35 million for 17 pilot projects to combat erosion, saltwater intrusion, subsidence, and wetland loss along Louisiana's coast. In 1989, the Louisiana legislature established the State's Coastal Wetlands Conservation and Restoration Trust Fund. Deposits in the trust fund are based on a percentage of the state's mineral revenues and have varied from \$13 million to \$25 million annually, depending on oil and gas prices and production levels. The trust fund has been the source of the state's cost-share for most coastal restoration projects undertaken by the Corps in Louisiana during the past decade.

The common understanding of the extent, causes, and impacts of wetland loss in Louisiana, coupled with the 30-year history of federal and state coastal protection laws and funding authorizations, has enhanced the coordination and integration of coastal protection programs among federal and state agencies in Louisiana. As early as 1980, the Corps' New Orleans District began utilizing the expertise of the U.S. Geological Survey and the U.S. Fish and Wildlife Service to provide habitat data and maps in support of project planning in the Louisiana coastal zone. In 1990, the New Orleans District designated a full time project manager for the environment to help coordinate work with resource agencies and academic institutions.

In 1990, the U.S. Congress passed the Coastal Wetlands Planning, Protection, and Restoration Act (CWPPRA) (P.L. 101-646,), locally known as the Breaux Act. In Louisiana, this act created a partnership between the

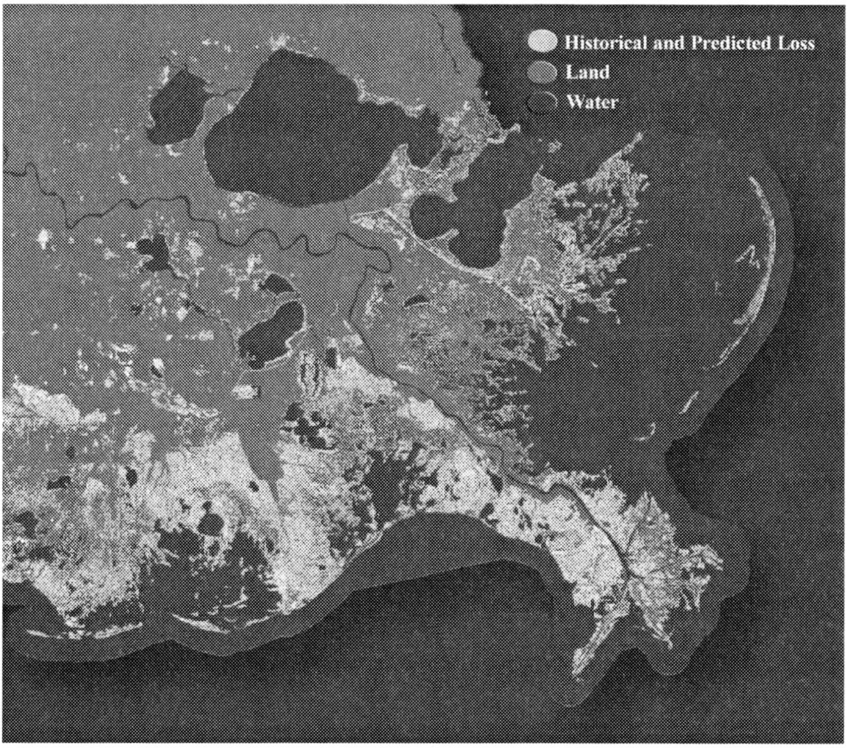

FIGURE 4-1 Historical loss depicted in this image includes data from 1932 to 2000, and projected loss is based projected trends from 2000 to 2050. Data from Barras et al. (2003); available [on-line] at *http:// www.coast2050.gov/images/landloss8X11.pdf.* Figure courtesy of John Barras, U.S. Geological Survey, National Wetlands Research Center, Baton Rouge, Louisiana.

state government and five federal entities—the U.S. Departments of Army, Agriculture, Commerce, and Interior and the U.S. Environmental Protection Agency—to implement wetland restoration projects with approximately $40 million per year of dedicated federal funds. The Corps chairs the Louisiana Coastal Wetlands Conservation and Restoration Task Force created by the Breaux Act. In 1993, the task force, as mandated by Congress, produced a multi-volume Coastal Restoration Plan that proposed numerous projects of varying scales. In addition, a total of 137 projects were authorized for Breaux Act funding. These relatively small-scale projects were estimated to preserve more than 100,000 acres of marsh over their 20-year project life.

To date, 52 projects have been built and are being monitored by the state with the advice of an academic team.

In 1996, it became apparent to the task force that the Breaux Act restoration projects and the two freshwater diversions would prevent only 22 percent of the projected future wetland loss. Under Corps leadership, the task force initiated the "Coast 2050" study and produced a report that presented restoration strategies across the coast and a general implementation plan (Coastal Wetlands Conservation and Restoration Task Force, 1998). The Coast 2050 plan was revised and updated and renamed the Louisiana Comprehensive Coastwide Ecosystem Restoration (LCA) Study. The State of Louisiana is the local sponsor of the feasibility study. For purposes of the feasibility study, the Louisiana coast was divided into four coastal regions that were further subdivided into nine coastal Louisiana hydrologic units. Priority of study in each region is determined, in part, by the rate of wetland loss within each coastal unit.

To guide the development of the LCA feasibility study, the Corps' New Orleans District has formed a collocated team of federal and state environmental scientists and engineers, which is housed full- or part-time at the district. The collocated team consists of 15 team members representing six federal and state agencies. The expertise of team members includes marsh ecology, fish and wildlife biology, water quality, desktop modeling, and hydrology. These team members are involved in the preparation of the endangered species assessment, the coastal zone consistency determination, and the essential fish habitat reports that must accompany the feasibility study. Hence, members of the collocated team are involved in reducing the potential conflicts among regulatory agencies as individual projects are planned and constructed.

A National Technical Review Committee of outside experts is responsible for reviewing the LCA study plan throughout its development. A portion of the funding allocated by the Corps and the state for the feasibility study has been transferred to academic institutions for modeling support. The local academic community has also been involved with the interagency team in the development of desktop and numerical modeling efforts. Ecological and hydrological simulation models will be used to formulate projects and to analyze alternatives. A monitoring plan will be developed for each project proposed in the LCA plan, which should enhance the potential for adaptive management and allow course corrections as projects are built and operated.

Integrated Planning and Sediment Management

Another example of the Corps' use of integrated planning in coastal systems and river basins can be seen in some of its sediment management practices. Because of the regional nature of water and sediment problems, the Corps has developed an integrated research and demonstration effort called the Regional Sediment Management Program (RSMP). The overall goal of the RSMP is "to ensure that water resources projects throughout a sediment region affect sediment, and are affected by it, in an economically and environmentally sustainable manner. It recognizes that the region and embedded ecosystems respond beyond the space and time scales of individual projects, and that a proactive regional planning and engineering approach can produce significant national benefits" (U.S. Army Corps of Engineers, 2002d). This program represents a significant investment by the Corps in developing and demonstrating tools and knowledge in order to manage sediment resources on a regional basis. The program was initially developed with a coastal focus and had demonstration projects in several districts (Mobile, Jacksonville, Philadelphia, New York, and Detroit), and in the South Pacific Division. The need for system planning is particularly evident in cases where river control works have interrupted the delivery of sediment to the coastal zone (see case study below on the California master plan for coastal sediment management). These reductions in coastal sediment supply, which can arise from channel stabilization and sedimentation in reservoirs and detention facilities, can lead directly to accelerated coastal erosion. Jurisdictional boundaries and administrative divisions within the Corps can work against a coordinated effort to develop a efficient regional solution, making integrated planning essential. The RSMP is now being expanded to include inland watersheds.

Case Study: California Master Plan for Coastal Sediment Management. California beaches not only represent an ideal for the state and the nation, but also provide protection of critical habitat and infrastructure, and generate recreation and tourism revenue measured in billions of dollars. Much of California's coast is actively eroding (Nichols, 2003), and an essential factor driving that erosion is a reduction of sand supply from coastal watersheds. Dams, debris basins, urban land cover, and river stabilization projects all contribute to a reduction in the supply of sediment to the coastline. Sustainable solutions to beach erosion problems are not found through shoreline protection or beach renourishment on a local scale, but require a regional approach that links sediment sources and pathways to beach erosion and deposition.

Recognizing the regional nature of California's beach erosion problem, a collaboration of federal, state, and local agencies, led by the Corps and the California Resources Agency, are developing the California Coastal Sediment Management Master Plan (California master plan) to evaluate California's coastal sediment management needs on a regional, system-wide basis. In addition to evaluating and prioritizing coastal sediment management needs, the California master plan is intended to identify approaches for restoring and maintaining coastal wetlands and beaches and to coordinate information on sediment sources and sinks and beach erosion. As part of this effort, in 1999, the Corps and the California Resources Agency initiated the California Coastal Sediment Management Workgroup (CSMW) to provide a working structure for moving beyond site-specific projects and toward regional solutions to beach erosion problems. The CSMW provides a forum for stakeholders, including government agencies, nonprofit organizations, and the public, and serves as a planning and management entity for regional projects.

Under the auspices of the RSMP, the Corps has located a demonstration program within its South Pacific Division to support development of the California master plan; to develop a statewide approach for solving the sediment problems of shorelines, coastal wetlands, and coastal watersheds; and to quantify regional sediment budgets on a statewide basis.

Integrated Planning and Dredged Material

A related area in which the benefits of system planning have been recognized and supporting authorizations exist concerns the beneficial use of dredged material. Dredged material can be used for beach nourishment or for coastal wetland creation or restoration (see case study on reconstruction of Poplar Island). Although the beneficial use of dredged material is increasing, the proportion of all dredged material used for this purpose is still small, and a number of factors work against consistent attempts to use dredged sediments beneficially. Based on Corps data on federally contracted dredging projects, 7 percent (17.5 million cubic yards) of the nearly 250 million cubic yards of dredged material was used beneficially in 1999. In 2001, the amount of dredged material used beneficially rose to 50 million cubic yards (24 percent) of the nearly 210 million cubic yards dredged. The increase in beneficial use of dredged material indicates that means are being found to support the additional cost of the beneficial use of dredged material, although the majority of dredged material is not being used for this purpose. Economic incentives are needed to keep the natural resource of dredged material within river and coastal systems so that the material is not lost to the environment permanently.

A dilemma for the Corps and its district offices is that typical dredging projects for navigation channel maintenance are selected on the basis of the least-cost alternative. Although specific authorities exist to allow selection of other alternatives when these alternatives can be shown to provide environmental benefit, short-term financial and scheduling pressures can work against their selection. In most cases, using dredged material beneficially will increase the cost substantially above the least-cost alternative. Because district operation and maintenance (O&M) budgets are sometimes insufficient to cover all the required dredging projects, increased costs associated with beneficial use may come at the expense of delaying other scheduled projects. If districts are proactive, however, they can request beneficial-use funds that are available in the current WRDA. However, the funds designated for beneficial uses in the WRDA are generally much less than what is needed, and some districts may not even apply for these funds because the chances of success are poor.

Districts face a similar dilemma in developing feeder berms that leave clean sand in the coastal environment or pumping sand to a confined disposal facility for storage. These options for managing dredged material are typically more costly than the least-cost alternative. The funds for such projects must come from current O&M funds, from Section 204 or Section 207 authorizations of the WRDA, or from a local sponsor. The least-cost alternative will continue to be attractive to project managers unless some incentive is provided to pursue beneficial-use alternatives. For example, when beneficial use of dredged material delays other projects, the district's budget in subsequent years could be increased by the amount equal to that spent on the beneficial use in order to complete the delayed projects. Another alternative would be to assess fees associated with channel deepening and maintenance and then use a portion of these fees to fund beneficial-use projects and thereby create greater incentive to undertake such projects.

Recommendation 4-4: Dredged material should be viewed as a natural resource and efforts should be directed at conserving this resource. A system is needed to determine the best use of the dredged material, rather than the least-cost alternative for its disposal. Sufficient funds and incentives should be provided through Sections 204 and 207 of

WRDA, by local sponsors, or by using innovative methods to further increase the beneficial use of dredged material.

Case Study: Reconstruction of Poplar Island. Poplar Island refers to a group of small islands that were originally a single island, located in the upper middle of the Chesapeake Bay. To reconstruct the island to its 1847 size, the Poplar Island Project uses dredged material from the Baltimore Harbor and Channels federal navigation project (Figure 4-2). The project allows for the disposal of a portion of the spoil from extensive maintenance dredging and shipping channel improvements and is expected to generate as much as 100 million cubic yards of dredged material over the next 20 years (U.S. Army Corps of Engineers, 2002e). The project was approved under Section 204 of the WRDA 1992 (P.L. 102-580), and funding was authorized under the WRDA of 1996 (P.L. 104-303). The goals of the project are to optimize the volumetric capacity of the site for dredged material, restore Poplar Island to its 1847 footprint, and create and restore habitat (Headland et al., 2000). The original project cost of $427 million has been reduced to $340 million because the amount of dredged material stored has been reduced from 38 million or 40 million cubic yards to 33 million cubic yards.

The plan is to use clean sediments dredged from the Chesapeake Bay navigation channels to create approximately 1,110 acres of wetland and upland habitats consisting of 111 acres of high marsh, 444 acres of low marsh, and 555 acres of upland habitat. The project is expected to restore the island, create wildlife habitat, and enhance the Chesapeake Bay through increased habitat diversity. The total project life is estimated at 24 years, with the placement of approximately 2 million cubic yards of dredged material per year.

Dikes for containing the dredged material were constructed in two phases. Phase 1 covered approximately 640 acres and included the construction of 25,000 feet of armored perimeter dike, more than 11,000 feet of unarmored interior dike, and a breakwater more than 2400 feet long. Placement of dredged material inside the diked area began in the summer of 2000. Phase II began in 2001 and included the construction of 15,000 feet of armored perimeter dike and approximately 8,000 feet of unarmored interior dike, and this phase covered about 470 acres. As the approximately 2 million cubic yards of dredged material is placed within the containment dikes each year, the elevation is expected to rise to the design elevation of 20 feet above mean low water. The deposited material will then be shaped to provide approximately 80 percent low marsh and 20 percent high marsh,

FIGURE 4-2 Poplar Island Environmental Restoration site located in mid-Chesapeake Bay. Photo courtesy of Scott Johnson, U.S. Army Corps of Engineers, Baltimore, Maryland.

with small upland islands, ponds, and channels to increase habitat diversity. Planned by the Corps in collaboration the U.S. Fish and Wildlife Service, the Maryland Department of Natural Resources, and other resource management agencies, the completed project is identified as valuable nesting and nursery area for many species of wildlife, including eagles, osprey, heron, and egret, and it will help to address the historical loss and degradation of this habitat in the Chesapeake Bay.

BARRIERS TO IMPLEMENTATION AND FACTORS FOR FUTURE SUCCESS

The most important factor determining whether an integrated watershed, coastal system, or ecosystem study is completed is the existence of a specific authority for the study and its associated funding. Specific authorization may be developed in response to a disaster such as a flood, to findings of jeopardy under the Endangered Species Act (particularly for anadromous fish; P.L. 93-205), or to significant degradation of a widely valued ecosystem resource (such as the Chesapeake Bay or the Florida Everglades). In other circumstances, authorization may be developed in response to local or state pressures that prompt congressional interest. Although programmatic authorities exist to support integrated systems studies, these studies have not historically had sufficient funding to allow widespread implementation. Further, such efforts have required special initiative on the part of project managers. In some cases, project managers' enthusiasm and resource-fulness allow them to take advantage of existing programmatic authority. In other cases, a lack of expertise, enthusiasm, or funding; competing pressures from local sponsors; or turnover in key personnel will act to prevent an integrated watershed or coastal system plan.

Many examples of integrated planning efforts have high stakes, leading to a focused and highly visible attempt to identify the salient issues, incorporate input from all relevant stakeholders, and balance competing objectives. Many Corps water projects have lower visibility. The panel sought to develop some indication of the extent to which ordinary water projects incorporate integrated systems planning. Although the panel received a range of comments regarding integrated systems planning efforts at the district level, these comments were—in the absence of a thorough audit beyond the scope of the panel's available resources—largely anecdotal. Thus, it was difficult to judge the proportion of Corps projects that effectively incorporate watershed or coastal system evaluations in their planning. Indeed, it was hard to even judge the proportion of Corps projects that claim to account for watershed or coastal system factors. Such studies may appear in different portions of the Corps' budget (e.g., general investigations, O&M, construction). Some projects may incorporate a significant systems component but have a title that provides no indication of this. Other projects may be titled as, for example, ecosystem restoration projects but may actually include only modest efforts in that regard. Thus, it is not possible to conclusively demonstrate the extent to which adequate systems analysis is carried out to support Corps projects.

A Corps study of the use of watershed planning analyzed 10 environmental restoration and flood damage reduction projects dated 1999-2001. Hansen and Fischenich (2002) developed a list of topics indicative of a watershed perspective in a planning study and evaluated whether or not these items were discussed in the planning documents. They found that the planning studies did a very good job in areas that have long been part of the Corps' planning methodology, such as characterizing existing conditions, formulating the problem, identifying alternative approaches, and evaluating social benefits and costs. Less consistency was found in evaluating hydraulics, hydrology, and sedimentation at the watershed scale; in defining cause-effect relations between physical and biological components of the watershed; and in incorporating adequate adaptive management plans. These are elements that focus on interactions that operate at broader scales than the immediate project area and, therefore, are a central part of integrated water resources planning within river basins and coastal systems.

Some of the most visible integrated water resource projects have a large spatial scale, address significant economic and environmental issues, and have a high regional and national profile. As for any large-scale water project with important environmental consequences and a range of stakeholders with conflicting interests, the quality or adequacy of the Corps' systems-level planning efforts has been debated in each of these cases. The varied, overlapping, and conflicting objectives in such projects prevent a final project plan that will fully satisfy all parties. In this context, the criteria for success include not just a balance of economic and environmental benefits that satisfies all interested parties, but also a broad-based effort to openly consider all significant issues, applying the best available science and engineering to develop objective information to support the policy choices to be made. The panel was able to find many examples of the Corps' success in the latter regard.

It is evident to this panel that the Corps, when given the authority and funding, is capable of multi-stakeholder, multi-objective planning projects that incorporate a diverse range of economic and environmental issues over the necessary spatial and temporal scales. Although it was not possible to prepare a "report card" quantitatively on the Corps' performance in using integrated water resources planning, it was nonetheless evident that its application was less consistent than desired and that particular deficiencies and barriers exist that, if addressed, would allow the Corps to perform more consistently and effectively.

5

Toward Integrated Water Planning and Management in a River Basin and Coastal System Context

Chapter Highlights

This chapter evaluates constraints that the Corps faces in fulfilling its mission to plan and evaluate water projects in an integrated river basin and coastal system context. It considers the existing water policy environment in the United States and the organizational, jurisdictional, and regulatory issues that can work against consistent implementation of integrated water resources planning and evaluation. Steps that Congress and the Corps can take to improve its performance are suggested.

THE CURRENT POLICY AND PROJECT ENVIRONMENT

As pointed out throughout the previous chapters, water resources planning and management in a river basin and coastal system context requires an integrated approach to provide a balanced consideration of objectives and potential impacts at relevant time and space scales. The need for such an approach is widely endorsed by the water resources planning and management community (NRC, 1999a,b). The Corps has embraced these principles and, in many cases, has played a major role in their definition and in the development of supporting methods. The Corps' adoption of integrated water resources planning and management is clearly stated in its Policy Guidance Letter (U.S. Army Corps of Engineers, 1999d):

> There is a growing recognition that "locally perceived water resources problems" have regional dimensions and are of concern to numerous, diverse interest groups. Many activities occurring in a watershed are interrelated and, therefore, managing water resources has evolved to more of a holistic, collaborative effort. The Corps has developed its own watershed perspective to guide water resources development, protection, and management within the Civil Works program. This watershed perspective accommodates the multi-objective, multi-purpose planning and investigations necessary for exploring these concerns.
>
> The watershed perspective applies to all Civil Works programs through planning, design, construction, operation, maintenance, restoration, rehabilitation, and regulatory activities. The application of this perspective into the Civil Works program encourages opportunities for enhancing the operations and maintenance of existing projects, especially the management of the natural resources.

In the same document, a watershed approach is stated as Corps policy:

> Policy: The Corps will integrate the watershed perspective into opportunities within, and among, Civil Works elements. Opportunities should be explored and identified where joint watershed resource management efforts can be pursued to improve the efficiency and effectiveness of the Civil Works Programs. The Corps will solicit participation from Federal, tribal, state, and local agencies, organizations, and the local community to ensure that their interests are considered in the formulation and implementation of the effort. Due to the complexity and interrelation of systems within a watershed, an array of

> technical experts, stakeholders, and decision-makers should be involved in the process. This involvement will provide a better understanding of the consequences of actions and activities and provide a mechanism for sound decision making when addressing the watershed resource needs, opportunities, conflicts, and trade-offs.

In this context, the term "watershed" is interpreted by the Corps and others to indicate not only terrestrial watersheds, but also coastal systems. This connection is made explicit in other Corps guidance and policy documents, such as those describing the Regional Sediment Management Program (e.g., Martin, 2002; U.S. Army Corps of Engineers, 2002d,f). The Corps' commitment to a watershed approach is also formalized in the *Unified Federal Policy for a Watershed Approach to Federal Land and Resource Management* (65 Fed. Reg. 62566, October 18, 2000), which was adopted by the Corps and other federal agencies.

Clear support for an integrated planning approach has also been provided by Corps leadership. In testimony before the U.S. Senate (U.S. Senate, Committee on Environment and Public Works, 2002), Chief of Engineers General Robert Flowers stated:

> Right now, existing laws and policies drive us to single focus, geographically limited projects where we have sponsors sharing in the cost of the study. The current approach narrows our ability to look comprehensively and sets up inter-basin disputes. It also leads to projects that solve one problem, but may inadvertently create others. Frequently we are choosing the economic solution over the environmental, when we can actually have both. I believe the future is to look at watersheds first; then design projects consistent with the more comprehensive approach.

As pointed out in Chapter 4, there are a number of authorities in place that may be used to undertake watershed studies, planning, and management. These include continuing and programmatic authorities that allow projects for the purposes of restoring and protecting ecological resources without specific congressional authorization (Table 4-1).

Despite existing policy and planning guidance and existing authorities, the Corps' record in integrated water planning and management is inconsistent. Important project successes occur hand-in-hand with projects that have unanticipated consequences, significant cumulative impacts, or undesirable economic and environmental outcomes. These undesirable impacts often make a project unacceptable by today's standards. To some extent, the panel acknowledges that these projects were designed to meet a suite of objectives that was narrower than the diverse social, economic, hydrologic, and ecologic objectives of today. Nonetheless, in the panel's discussions with Corps planners, it was clear that various institutional, jurisdictional, and knowledge barriers can stand in the way of implementing the broader mandate of integrated water resources planning and management.

NATIONAL WATER POLICY

Until the mid-1970s, national water policy was premised on the need to alter the hydrology of river and coastal systems with dams, levees, dredging, channelization, and other physical improvements to "optimize" the economic benefits of water use. As a result of the environmental movement and persistent criticism of the economic benefits of many large water-related public works projects, the era of large-scale dam building (and similar projects) in the United States came to an end. No clear new paradigm, however, has emerged to replace the old multi-objective development model. Today, traditional water constituencies compete primarily with environmental interests for control of management of the nation's waters. In the absence of clear guiding authorities, competing interests are resolved in an ad hoc fashion through congressional action or stakeholder processes. This policy landscape severely reduces the nation's ability to plan and manage effective water resource projects using a systems approach.

Today, water resource planning in the United States may best be described as guided by a de facto national water policy that is enunciated in legislation, administrative policy, and court decisions. Nationally, there is a lack of an institutional instrument to set policy at a river basin or coastal system level within which large and small water resources projects can be developed and evaluated (Kenney, 1997; Loucks, 2003; NRC, 1999a, b; Stakhiv, 2003). This lack of organizational framework in water resources planning and management has hampered the Corps' ability to consistently plan water resources projects within an integrated systems context.

At the center of this de facto policy are a number of environmental laws (e.g., the National Environmental Policy Act of 1969, the Clean Water Act of 1972, the Endangered Species Act of 1973). Not only do these laws redistribute responsibility for water policy among different federal agencies, they also create venues for private citizens and nongovernmental organizations (NGOs) to influence environmental policy through the courts. The sum of these actions, including case law, provides guidance and constraints for water project planning and increases the emphasis placed on environmental objectives along with the traditional economic and commercial purposes of federal water resource projects.

As discussed in Chapters 2 and 4, an almost unavoidable consequence of relying on such a decentralized framework is that the policies, regulations, and case law that make up the de facto policy are sometimes vague and often in conflict. The divided authority over water quality provides one example. Water releases from Corps and other dams play a role in water quality maintenance, but water quality standards are set by the each state, with oversight provided by U.S. Environmental Protection Agency (USEPA). This arrangement creates gaps that must be addressed by federal and state agencies with water resources management missions. Another example of recent prominence (summer of 2003) concerns dam releases on the Missouri River. In this case, the Corps job was complicated by two binding and conflicting federal court decisions, one requiring that water be released to maintain a 9-foot navigation channel and the other requiring that releases be decreased to enhance habitat for endangered species.

Such conflicts provide good examples of the negative consequences of the decentralized and conflicting guidance for water resources management, but do not illustrate the opportunities for wise water management that may be missed in such a planning environment. Ideally, there should exist a set of clear, internally consistent water resource objectives, applied through master plans for river basins and coastal systems. These objectives should delegate clear lines of authority and funding for implementing the plans. Although such a structure would undoubtedly make the job of integrated water planning and management easier, at this time the Corps will have to continue operating without such coordination.

The relative importance of the wide range of water resource objectives—hydrologic, geomorphic, ecologic, social, and economic—will vary from one region to another and from one time to another. The number and complexity of water resource objectives, particularly for studies integrating economic and environmental objectives over a range of spatial and temporal scales, shift the focus from directed, optimized solutions to trade-offs among competing objectives. Decisions about which objectives to optimize cannot be based exclusively on an analytical methodology, particularly when considerable uncertainty is associated with any forecasts of a project's benefits and impacts. Rather, value-based judgments are required of those with decision-making authority. The societal values on which these judgments are based are not likely to remain constant over time or from region to region. A good example is recognition of the importance of environmental values in project planning over the past four decades. The increasing role in decision making assigned to state and local entities makes regional variations in project priorities inevitable. In this context, the Corps' primary mandate should be to apply the best available science, engineering, and management tools toward developing the information needed to support the decisions required of local, state, and federal entities charged with the authority for making those decisions. When conflicts exist between different federal agencies (e.g., between the Corps and the National Marine Fisheries Service [NMFS] or the U.S. Fish and Wildlife Service [USFWS] regarding endangered species habitat, or between the Corps and USEPA regarding clean water actions), adjudication within the executive branch of the federal government is possible.

PIECEMEAL APPROACH TO WATER PROJECT PLANNING

Currently, water projects are approved by Congress on a case-by-case basis. This present project environment, in which a collaboration of local entities and Congress can exert considerable pressure for a quick and favorable project evaluation, does not support a balanced and integrated evaluation of all project benefits and costs at all necessary time and space scales. This pressure is cited as the primary barrier to more integrated planning by Corps leadership (U.S. Senate, Committee on Environment and Public Works, 2002), Corps staff, and the press, particularly in the context of controversial, often legacy, projects. Although there are many cases in which the Corps conducts or contributes to a comprehensive study (particularly when such a study is the specified objective of the project), there are many other cases in which immediate economic concerns regarding navigation, flood protection, or erosion control exert strong pressure on Corps managers—pressure that often works against a balanced consideration of important project benefits and costs, particularly those related to environmental resources that are difficult to quantify.

Ultimately, Congress and the Executive Branch must choose whether the broader goals of integrated water planning take priority over the immediate benefits of a particular project. The water resources science and engineering community can provide an understanding of the relevant hydrologic, geomorphic, ecologic, social, and economic issues involved, along with decision support tools and a sense of diverse stakeholder views on particular projects. It cannot ensure that those with decision-making authority make prudent or popular decisions. Therefore, recommendations provided in this report focus on a means by which the Corps can do a better job in providing the best possible scientific and engineering evaluation, in collaboration with other government agencies and relevant stakeholders.

CONSTRAINTS ON PROJECT PLANNING AND EVALUATION

In the current Corps environment, integrated planning often occurs in response to specific circumstances. Authorization may develop in response to a disaster such as a flood (e.g., the Sacramento-San Joaquin project), to jeopardy findings under the Endangered Species Act (P.L. 97-304) (particularly for anadromous fish), or to significant degradation of a widely valued ecosystem resource (such as the Chesapeake Bay or the Florida Everglades). Alternatively, authorization may develop in response to local or state pressures that prompt congressional interest.

Integrated water resources planning and management requires adequate consideration of all relevant objectives and all potential impacts—local, regional, and cumulative. The time and space scales for such studies cannot be specified in advance, and no simple recipe exists for defining them. Flexibility is needed in defining the appropriate scope of evaluation studies. Yet reconnaissance studies are limited by law to a budget of $100,000 and to a period of 12 months (18 months in some cases). This is a severe restriction that makes it extraordinarily difficult to perform an adequate system evaluation except for small, simple projects.

Current limitations on reconnaissance studies may be adequate for some projects with an obviously narrow scope but will clearly be inadequate for many others. Removing time and budget constraints may have the undesired consequence of allowing persistent growth of studies—leading to an agency dominated by study gridlock and few useful projects. This criticism has been raised in the past and has led to various proposals for reforming Corps planning methods (NRC, 1999b). Nonetheless, the alternative to specifying constraints for all projects is an inefficient solution that will inappropriately curtail a truly necessary study while encouraging the expansion of minor studies to the specified limits. Support for a more flexible approach can be found in a survey exercise conducted among Corps planning staff at their 2002 meeting in New Orleans.

Recommendation 5-1: The Corps should develop a rigorous, in-house proposal and review process in which the scope and budget of reconnaissance studies are determined on a case-by-case basis.

Another impediment on the Corps' ability to undertake project planning at broader spatial and temporal scales is the requirement for local cost-sharing enacted in the Water Resources Development Act (WRDA) 1986 (P.L. 99-662), which has resulted in reducing federal share and increasing the non-federal share of costs in federal water projects. The introduction of greater local cost-sharing has impacted the Corps in positive and predictable ways, but also in unanticipated negative ways. Although non-federal beneficiaries of Corps projects have long borne some of the costs on an ad hoc basis, mainly in the form of land and easement transfers and dredged material disposal areas, momentum for increased fiscal discipline in the process began to build in the 1970s. The Corps had long been criticized for flawed benefit-cost studies, especially for overstated benefits. A coalition of fiscal conservatives and environmentalists agreed that formal cost-sharing was a desirable Corps reform because it would eliminate projects of marginal value and require that the community—via a local sponsor—have more of a vested interest in the project.

Prior to the WRDA 1986, there was little incentive for local, non-federal parties to reject any project that was proposed. These local parties could expect to receive the bulk of project benefits, while being responsible for only a small part of project costs. For example, prior to WRDA 1986 there was no cost-sharing requirement for structural flood control projects. After enactment of WRDA 1986, nonfederal parties were required to share between 25 percent and 50 percent of costs, including a minimum 5 percent required contribution in cash—as distinct from in-kind contributions such as easements, rights-of-way, and land to dispose of dredge material (DelRossi and Inman, 1999). A principal reason for changing pre-WRDA 1986 cost-sharing rules was to increase the likelihood of federal tax dollars being spent on economically and socially worthwhile water projects. Advocates for this change in WRDA 1986 argued that without more extensive cost-sharing requirements, the process of planning and executing federal water projects was biased in favor of approving many less-than-worthwhile projects.

There is general agreement in the water resources community that the 1986 WRDA has significantly affected both the process by which water projects undertaken by the Corps are planned and executed and the nature and scope of projects that are funded. A careful empirical study of water project selection pre- and post-WRDA 1986 indicates that increased local cost-sharing resulted in a downsizing of projects ultimately requested and approved in consultation with local sponsors (DelRossi and Inman, 1999).

As a result of the 50 percent cost-sharing requirements between the Corps and non-federal sponsors in the feasibility stage of most projects, local communities have become much more involved in planning. This creates a very powerful incentive for the Corps to acquire non-federal partners that will agree to cost-share planning. In fact, it is mandatory for the Corps to have already identified a prospective cost-sharing sponsor during the reconnaissance phase, and the feasibility phase cannot begin until a local sponsor signs a contract or feasibility cost-sharing agreement

A negative effect of the shift toward more local cost-sharing has been the creation of subtle, but real, incentives for the Corps to focus on local projects in the absence of any river basin or coastal system master plan. Since a typical non-federal sponsor may have little incentive or legal ability to consider the environmental, economic, and hydrological effects of such projects beyond those experienced locally, the post-WRDA 1986 cost-sharing rules favor single-purpose projects with well-defined local benefits because local sponsors tend to promote investments that address specific local needs. As a consequence, the Corps may have limited opportunity to consider projects that involve planning at a broader environmental and hydrological scale than that desired by the local sponsor (NRC, 1999b).

Thus, although there is movement by the water resources community toward broader spatial and temporal planning on a watershed level (NRC, 1999a, 1999b; Schad, 1998), and even though the *Economic and Environmental Principles and Guidelines for Water and Related Land Resources Implementation Studies* (*P&G*) states that "civil works planning should incorporate a watershed perspective . . . " (WRC, 1983, Section 2-8), the WRDA 1986 cost-sharing rules have worked at cross-purposes, albeit unintentionally, by reducing the opportunity for the Corps to plan and execute projects at a spatial scale that is consistent with watershed and coastal systems planning. Indeed, when the Corps has undertaken projects at a watershed scale, it is often because the "political boundaries" of local sponsors roughly matched those of particular watersheds. For example, both the Sacramento-San Joaquin and the Everglades projects concern watersheds that lie entirely within a single state.

The simple solution may seem to be to a return to pre-WRDA 1986 cost-sharing arrangements. However, this alternative ignores the fact that cost-sharing was put in place in 1986 to remedy some real problems in the Corps' operations. Among these were (1) a tendency for local interests to support federal spending on water projects because these appeared to be "free" (or heavily subsidized) without cost-sharing and (2) a tendency for planning of Corps projects to take a great deal of time in the absence of pressure from local sponsors to "get the plan moving."

A more appropriate remedy would be to devise cost-sharing arrangements that strike a better balance between holding local interests accountable and planning on a broader spatial scale. An important step in this direction would be to recognize that while it remains appropriate to hold local sponsors accountable for costs associated with projects that are truly local in scope, or for aspects of projects with well-defined local benefits, local sponsors also have little interest, authority, or ability to support integrated studies over broader regions. In contrast, a consistent and integrated evaluation of all objectives at all relevant time and space scales is clearly a federal interest. For this reason, integrated evaluation of project benefits and costs at the river basin and coastal system scales should be federally funded.

Recommendation 5-2: Reconnaissance studies should include an integrated evaluation of all project benefits and impacts at any relevant spatial and temporal scale, leading to a definition of the scope and budget of integrated river basin or coastal system analyses required in a feasibility study. Cost-share requirements for feasibility studies should be amended to provide 100 percent federal funding for an integrated evaluation of project benefits and costs at all relevant temporal and spatial scales.

In the reconnaissance phase, the panel recommends the incorporation of an integrated analysis to provide the necessary evaluation of all appropriate benefits and costs at all relevant scales of space and time. Advances in information and decision support technology can provide considerable assistance during this analysis within the reconnaissance study framework. In addition to the current requirement of developing a cost-sharing agreement with the local sponsor, the panel proposes that requirements for a reconnaissance study be amended to include the definition of those portions of the feasibility study that constitute integrated river basin or coastal system evaluation. The panel also proposes that these integrated evaluations in the feasibility phase be 100 percent federally funded because they would assess the portions of the project concerned with a broader evaluation of benefits and costs. However, the panel emphasizes the need to preserve the existing 50 percent cost-share requirement for those portions of the project directly concerned with project development (i.e., design, land acquisition, construction).

Although local sponsors may still be eager for an early and positive finding to proceed with their project and may exert pressure to curtail an integrated evaluation, a revision of the funding formula for feasibility studies ensures that an unwillingness or inability to pay for the study will not prevent the necessary integrated evaluation from going forward. A clear mandate for the Corps to pursue such studies would assist Corps planners in following the federal mission. This mandate can effectively be addressed by specifying the need for an integrated evaluation within the specific requirements of a reconnaissance study.

Recommendation 5-3: The scope and budget for integrated planning studies should be determined in the reconnaissance phase and explicitly defined in the project study plan and cost-sharing agreement that define the scope and financial responsibilities of the feasibility study. Approval of a feasibility study should be contingent on a judgment, informed by appropriate internal and external review, that a study plan of the salient social, economic, and environmental factors, at all relevant spatial and temporal scales, has been defined.

CONSTRAINTS ON COLLABORATION

The Corps is one of many federal agencies with responsibilities for water resources planning and management. Significant land management and water pollution authorities reside with the states. Stakeholder participation is now widely accepted and actively endorsed by the Corps. The environment in which the Corps now evaluates, develops, and operates water projects is complex and includes many participants. Both the complexity of the issues considered and the number of the stakeholders involved increase when the focus of the investigation is at the scale of river basins and coastal systems, where there is a complex and conflicting mix of objectives inherent in evaluating water resources projects. In this environment, collaboration in data gathering, analysis, alternatives development, and decision making is essential for effective water management.

In addition, implementation of federal water policy is decentralized, with different agencies responsible for different aspects of an ecosystem (e.g., the Bureau of Land Management, USEPA, and USFWS all make decisions that affect stream water quality and use), including those ecosystems experiencing degraded environmental quality (e.g., reduced fisheries yields, poor water quality) due to natural and human-influenced processes. Adopting a system perspective in the planning and operation of federal water projects recognizes that these various subsystems function as whole, integrated units—not as separate parts. Adopting a systems and watershed perspective, therefore, requires improved coordination among agencies, as well as the funding and training required for this coordination.

Watershed planning requires cooperation with multiple agencies at federal, state, and local levels of government, as well as organized groups of stakeholders. The panel received conflicting reports of the Corps' ability to engage in formal collaboration with other federal agencies. Some reported that interagency collaboration was feasible and even routine. Others reported that the Corps' funding authority has limited its partnering with other federal agencies, particularly in cases where it is not the lead agency. A review and standardization of procedures for developing a memorandum of understanding (MOU) or memorandum of agreement (MOA) with other agencies is warranted, particularly when these other agencies have controlling authority (e.g., NMFS or USFWS in cases involving endangered species). Improved interagency collaboration is needed to appropriately address the ecological impacts of multiple anthropogenic stressors influenced by Corps projects and permitting and to promote the ecological and economic sustainability of our natural resources.

Some of the differences in the ability to collaborate with other agencies may be based on differences in the way each Corps district office manages its own affairs. District office autonomy was an intended feature when Corps districts were defined so that each office would have the latitude to deal with local matters of which it would have a better understanding than the regional office or Corps Headquarters. In general, this is appropriate and should not be changed. One result, however, is the appearance of conflicting operating procedures between districts, limiting true system planning and implementation.

Recommendation 5-4: Congress and the Executive Branch should take the steps necessary, including a standardized procedure for cooperative agreements and MOUs or MOAs, to ensure that the Corps is able to work effectively in collaborative planning and management.

The Corps is not the only entity with an interest or authority in water planning. The need to integrate ecological and socioeconomic issues with traditional water resource objectives across many federal and state agencies increases the complexity of project planning and admits a much larger number of agencies and stakeholders to the deliberations.

As environmental stewardship becomes more integrated into project planning, construction, and operation, the Corps must take into account that environmental consequences of project development may occur outside the watershed in which any particular project resides. No clearer example of this can be given than the Mississippi delta and Louisiana coastal system. While the Corps' New Orleans District has jurisdiction over the Mississippi delta and Louisiana coastal system, it has absolutely no jurisdiction in upstream watersheds where much of Louisiana's land loss and Gulf of Mexico hypoxia originate. The project alternatives that can be explored are only those that can be developed within the boundaries of the Corps New Orleans District.

As another example, dam construction that removes sediment from a river and eventually results in beach starvation at the mouth may not be an issue for a single Corps district whose geographical boundaries do not include both the coast *and* the watershed in which the dam is located. Similar problems arise from insufficient coordination between projects involving dredge material disposal in one district and beach nourishment in an adjacent district. There is no mandate for Corps' districts to cooperate; each district need only view its project independently, even though its project might be in close geographical proximity to another district's projects. Prioritizing dredging operations along the Atlantic or Pacific coast may include operations in multiple Corps districts. In effect, Corps districts compete for projects in Congress, without regard for the efficacy of the individual project within the sum of projects proposed in the coastal system. Although some attempts to coordinate sediment management have been initiated by the Corps, greater effort is needed.

KNOWLEDGE AND GUIDANCE ISSUES

State of the Art

State-of-the-art water resources systems planning requires abundant and diverse information that is organized into a useful framework for decision making. Technology to support effective decision making and planning is now widely available. The Corps' Institute for Water Resources (IWR), established in 1969 in Alexandria, Virginia, provides software to Corps districts for water resources planning and hydrologic engineering tasks. In 2000, the Corps' Navigation Data Center in Alexandria, Virginia, and the Corps' Hydrologic Engineering Center in Davis, California, were added to the IWR. The Navigation Data Center is the Corps' data collection organization for information on waterborne commerce, vessel characteristics, port facilities, dredging, and navigation locks. The Hydrologic

Engineering Center (HEC) is a research, training, and consulting organization that specializes in hydrologic engineering and hydrologic models used by Corps staff. The HEC concentrates on predicting and modeling river hydrology and hydraulics using various HEC models. These predictions are revised as the models become more refined and specialized (for more information, visit *http://www.iwr.usace.army.mil/iwr/software/software.htm*). In sum, the data and models provided and maintained by the IWR and its centers focus on hydrology, river hydraulics and sediment transport, hydrologic statistics and risk analysis, reservoir system analysis, planning analysis, and water control management. The Corps has a long and distinguished history in developing hydrologic and hydraulic modeling and decision support tools, and it has worked continually to develop more sophisticated and accurate modeling capabilities and will continue to do so in the future.

Although not maintained or served by the Corps, land-use and population models are now also widely available, as are remote sensing imagery and census data used to develop these models. Geographic information systems are routinely used to organize, analyze, and model relevant spatial information. The Corps' existing project scoping process, including public comment and consultation with other agencies, should be sufficient to identify and utilize existing land-use and population data and models.

The increased availability of data, software, and models does not mean that a complete planning support model can be developed in a short time with limited effort and expense, although it may be possible to conduct preliminary or screening analyses of considerable breadth and detail with modest effort. Such analyses can be particularly useful in a reconnaissance study to identify information needs that must be addressed in a more complete planning and decision-making effort. Some information needs may be satisfied by existing technology and resources for data collection (e.g., water quality samples, economic inventory) and modeling (e.g., refined hydraulic and sedimentation models), although the basis for fulfilling other information needs remains rudimentary (e.g., biological response to water projects).

The Corps has relied on hydrologic models in the design and management of water resources projects for decades, and few would argue that this is done poorly. However, important knowledge gaps and data needs exist. Water resources projects would be better served with more complete information in the following areas.

- ***Water data***. Long-term water discharge information is essential for evaluating the status of water resources and, in particular, any trends that can influence long-term, systems-based water resource planning. The number of water gauges used by the U.S. Geological Survey (USGS) is decreasing, and some long-term records are being terminated. Unreliable funding and decommissioning of USGS gauges is a serious problem for water planning, particularly when long-running records are terminated (Hirsch and Norris, 2001).

- ***Ecosystem knowledge***. Many Corps projects are designed entirely or in part to restore an ecosystem. The technical basis for modeling ecosystem response is still developing and often not sufficient to support reliable predictions. Often, general biological objectives can be specified only in terms of related ecosystem components, such as habitat restoration. Further research is needed to accurately assess ecosystem health and to positively connect water project attributes with ecosystem changes. Ecosystem models that predict the response of vegetation, changes in water quality, sediment delivery, and other ecological features (including surface water flows) to environmental change are being developed by some Corps districts. Habitat modeling is a core component of the Corps' adaptive restoration plans in the Florida Everglades and the Louisiana coastal area, but most Corps districts do not have access to the data or models needed to analyze the habitat impacts of water resources projects. Moreover, data sets needed to simulate environmental change (beyond hydrology) are woefully incomplete. Existing planning models and data bases are too simplistic to rely on for higher-order predictions of system and subsystem interactions. The ecosystem aspects of water project planning must account for such knowledge gaps, particularly by incorporating an adaptive framework with well-defined biologic targets that can be used to redirect management actions.
- ***Environmental economics knowledge***. Most Corps projects involve multiple and typically conflicting objectives. Effective water resource decision making on river basin and coastal system scale will require an improved ability to evaluate trade-offs among noncommensurate objectives. Methods for assessing environmental effects that can be implemented within existing Corps budgetary constraints and organizational structures must be developed. Although the Corps has taken steps to incorporate environmental effects in a systematic way in the planning process, it stops short of translating such effects into project benefits and costs that can be directly compared with national economic development (NED)

benefits and costs. This topic is discussed in more detail later in this chapter, as well as in Chapter 4.

Adaptive Management and Project Evaluations

The combination of multiple objectives, data gaps, knowledge gaps, and limitations in the state of the art introduces considerable uncertainty in project planning and design. This uncertainty is an inherent part of the management of all natural systems and its consequences are particularly obvious when ecological outcomes are added to the list of project objectives. To accomplish resource planning and management in the face of such uncertainty, the concept of adaptive management has gained increased acceptance in the Corps and other land resources agencies. As discussed extensively in the recently released National Research Council (NRC, 2004a) report *Adaptive Management for Water Resources Project Planning*, the concept of adaptive management is not new, nor is it complicated. Essentially it is a management tool that is multi-disciplinary and especially useful in the natural sciences.

Adaptive management is premised on the fact that neither natural nor social systems behave in a predictable, linear fashion, but rather are comprised of a significant amount of noise and inherent unpredictability. This unpredictability, or uncertainty, is particularly evident when various systems interact with one another. As a greater number of systems interact, the uncertainty increases with regard to the ability to predict an outcome. This uncertainty is amplified when applied to multiple, complex, and interacting systems (e.g., ecological response to changes in the hydrology, land use, and water quality of a watershed). Thus, predicting the outcome of a project that attempts to manage many complex systems is filled with varying levels of uncertainty.

In response to this uncertainty, it is useful to treat a management action as a type of experiment, whose results dictate how to continue with the project. In this context, it is necessary to identify key elements of the system whose monitoring will indicate the success of the project in meeting its objectives. Persistent monitoring provides the opportunity to change project features in ways that can correct for unintended or inferior results. Ongoing evaluations of project performance are critically important when dealing with increasingly complex and highly interactive systems.

Evaluations are a basic and integral part of the scientific process. Without some type of appraisal during or after construction of a project, it is impossible to know if the project succeeded and to identify parts of the project that require modification. Importantly, without post-project evaluation, it is not possible to learn how to improve the design and implementation of future projects. A post-project evaluation (PPE) is a type of summative evaluation conducted after the project is completed that focuses on the outcome to determine whether the project met the goals outlined at its beginning. For some projects, such as dam reoperation or beach nourishment, the necessary evaluation is ongoing and assesses whether the project is progressing as intended and, if necessary, informs changes in project design. Such evaluations have gained favor over the last decade as more Corps projects require performance data for adaptive management.

The PPE should not be limited to ecologic assessment but should examine all aspects of the project—technical, institutional, financial, social, management, and ecological. The PPE should reflect, at the very least, the same criteria with which alternative plans were reviewed in the feasibility phase. These criteria could be modified for a PPE in the following manner (U.S. Army Corps of Engineers, 1997):

- **Completeness**: Were the desired results obtained? Did the plan include all the necessary parts and actions to produce the desired results?
- **Effectiveness**: Did the project meet the plan's objectives?
- **Efficiency**: Did the project minimize costs and was it cost-effective?
- **Acceptability**: Is the project acceptable and compatible with existing laws and policies?

These criteria are a general guide for a project evaluation, whether the evaluation is carried out dung the project (ongoing) or after the project is completed as a PPE. Determining whether the project is "acceptable" requires that a determination be made at the beginning of the project to define a set of multiple and compatible objectives. As the project progresses toward completion, all of the above criteria can be used to ensure that the project achieves its goals in an effective and efficient manner.

The PPE does, however, require flexibility in the scope of the evaluation (not all projects would require an extensive evaluation) and should be explicitly defined, and cost-sharing agreed to, in the feasibility phase. Because the complexity and potential consequences will vary from project to project, the current cost limits on post-project evaluations should be replaced with a flexible system in which the scope, tasks, standards, and costs of PPE are determined on a case-by-case basis as part of a feasibility study. The decision to move ahead with a project should be contingent on the judgment, informed by appropriate internal and external review, that the post-project evaluation plan is sufficient to document the achievement of project objectives, as well as identifying unintended consequences and undesired cumulative effects associated with the project.

Recommendation 5-5: Post-project evaluations should be a required component of *all* projects and should be cost-shared with the local sponsor. The scope, timing, spatial and temporal scale, and funding for these evaluations should be determined during the feasibility study.

Further review of Corps evaluation and adaptive management procedures can be found in the reports on peer review, project planning, and adaptive management that form the other parts of this NRC review of Army Corps project planning methods and procedures (NRC, 2002a, 2004a, 2004b).

The Corps' Knowledge Base

The current state of knowledge is so vast in the water resources area that it can be difficult for one agency to harbor the full range of scientific, engineering, and socioeconomic skills that might be required on a particular project. The Corps must seek to find a useful balance of in-house and outside expertise. However difficult it is to build and maintain Corps expertise for these disciplines, it must be stressed that Corps management must continue to make a concerted effort to recruit, train, and maintain such staff. As Corps projects expand (e.g., more watershed analysis and planning) and diversify (e.g., more ecosystem restorations), they will have to rely on these subject matter experts to bring relevant professional experience and institutional knowledge for analyses and planning. Additionally, the Corps should take advantage of outside scientists and engineers who can bring specialized knowledge or a detailed understanding of the project area (in all aspects of the project's domain including economics, hydrology, etc.) These experts may be especially useful in filling specific knowledge gaps and possibly providing proprietary or other data sets that could help the Corps expedite and improve its planning. Outside experts in various disciplines could also help the Corps develop core competencies, whether they are hired as advisers or as contractors to train Corps staff.

The Corps, like other federal agencies, is currently losing significant parts of its institutional and core competency to retirement. An August 2002 memo on planning capability from Chief of Engineers George Griffin to division chiefs addresses the continued retirement of Corps personnel who were responsible for watershed-wide planning prior to the WRDA 1986. As these employees retire, they take with them the technical skills and institutional knowledge of how to conduct river basin planning acquired before the Corps moved more toward local, cost-shared, project-by-project planning. The Corps should consider conducting workshops and training seminars using senior Corps planners, involved in earlier water resources planning projects, to train junior planners in watershed-level water resources planning.

The Corps has recognized that the lack of training curricula has led to a knowledge gap in Corps planning and to a large number of planners with limited experience. As a result, the Corps is reemphasizing planning as an area of expertise and has instituted the following training initiatives: (1) planner core curriculum, (2) planning associates program, and (3) master's degree program. These programs are vitally important and should be consistently supported to provide not only effective in-house capabilities, but also incentive for the retention of strong employees.

Recommendation 5-6: The Corps should undertake an effort to review current staffing practices and, if necessary, expand these practices to maintain a well-trained in-house staff and to employ the services of outside scientists and engineers who can bring specialized knowledge or a detailed understanding of the project area.

Need for Revised Planning Guidelines

Fully integrated water resource planning and management requires effective guidance to determine appropriate time and space scales, and to evaluate non-commensurate objectives. Comprehensive guidance on integrated planning is not found in the current *Principles and Guidelines* (*P&G*), particularly regarding the evaluation of noncommensurate social, environmental, and economic objectives and the identification of appropriate spatial and temporal scales to analyze a diverse range of project objectives. Existing guidance is thorough on traditional benefit-cost analysis (BCA), but the heavy reliance on analytical methods must be relaxed in the context of multi-objective, multi-stakeholder integrated studies. The *P&G* has not been revised for 20 years and should be updated to provide sufficient and balanced information on how to conduct integrated water systems planning within river basins and coastal systems.

In today's environment, planning water projects at the river basin and coastal systems scale requires a balanced evaluation of environmental, social, and economic effects of water projects in project planning. For many years, a bedrock of the Corps planning process has been the use of BCA to assess the economic benefits and costs of water projects. Although the *P&G* recognizes the need to account for such effects through the national ecosystem restoration (NER) accounts, the challenge of monetizing environmental costs and benefits of water projects has prevented the Corps from weighing these costs and benefits directly along with the traditional economic benefits identified for inclusion in the NED account.

Methods for a balanced evaluation of environmental and economic objectives are evolving. Different approaches have been attempted. In some cases, environmental restoration has been included as part of the total NED cost of each project alternative (Box 4-1), as in the case of alternatives considered under the Sacramento-San Joaquin Watershed plan. In other plans, such as the Central and Southern Florida Indian River Lagoon South Project, environmental and economic benefits and costs are evaluated through the use of cost-effectiveness and incremental cost analysis. Under this procedure, the ecosystem benefits of alternative water projects are measured in terms of physical dimensions such as acres of wetlands or habitat units. Estimates are then made of the economic costs of achieving these physical changes in output, with the objective of determining the least-cost means of achieving each possible level of environmental benefit, measured in physical units. The resulting lists of "least-cost plans" are then compared to each other to identify the plans that produce more environmental benefits at the same cost or lesser cost than other alternatives.

These approaches to accounting for both NER and NED effects are among the feasible current alternatives, given continuing controversy over the reliability and validity of some approaches for monetizing environmental benefits and costs, but they can also exclude otherwise superior alternatives. Including the costs of environmental restoration while failing to include the economic benefits can cause some alternatives to appear costlier to society than is actually the case. Similarly, even sophisticated cost-effectiveness studies depend heavily on how the physical attributes of environmental benefits are defined and combined. The Corps should continue to take steps such as those outlined in *New Directions in Water Resources Planning for the U.S. Army Corps of Engineers* (NRC, 1999b) and *Analytical Methods and Approaches for Water Resources Project Planning* (NRC, 2004b) to develop protocols and standards for incorporating environmental benefits and costs into project planning in a manner that is comparable to traditional NED benefits and costs.

Recommendation 5-7: The Corps' planning guidance should be modified to provide Corps planners with contemporary analytical techniques necessary for integrated systems planning on large scales within river basin and coastal systems.

Guidelines for identifying all relevant factors affected by a water project and their spatial and temporal scales, and standards for a balanced evaluation of economic, social, and environmental factors, should be updated and expanded to a level of detail comparable to current standards for traditional BCA of the economic objectives of a project.

CONCLUSIONS

An ideal water planning environment—or even a reasonably good one—will take time to develop. The approach emerging over the past decade is one that emphasizes multiple objectives and gives a voice to a wide range of stakeholders. As the list of project objectives expands to more consistently include environmental and recreational objectives, the planning environment in which the Corps operates has become increasingly complex. The range of objectives and of the stakeholders who advocate them push water project planning and evaluation ever further in the direction of integrated analyses at the scale of river basins and coastal systems.

The Corps has demonstrated that it is able to participate effectively in integrated studies at the river basin and coastal system scale. Despite the need for integrated planning, the political and popular support for river basin and coastal systems planning has been neither consistent nor unanimous. Of particular importance is the powerful combination of a focused local sponsor with a well-defined project and a receptive member of Congress willing to carry the project forward to congressional authorization. The current piecemeal approach to project planning and evaluation works at cross-purposes to integrated water resources management at the river basin and coastal system scale. Such planning would clearly be facilitated within an entirely new context in which a clearly stated federal water policy was implemented by a central entity with the resources and authority to carry out the necessary analyses and select the most beneficial projects. The history of such central resource planning and management in the United States is mixed, and its prospects in the current fiscal and political climate are not strong. Nonetheless, the panel has identified interim, largely internal, steps that can be taken, although falling well short of comprehensive water systems planning. The actions proposed in this report would represent significant improvements within the Corps' planning environment that would raise Corps planning studies closer to the standards of integration and environmental stewardship articulated by federal regulations and Corps leadership.

Effective integrated water resources planning at the scale of river basins and coastal systems requires a clear mandate, consistent guidance and standards, and capable staff who are given the opportunity to evaluate all relevant aspects of a water project with adequate data, current tools, and necessary collaboration. The panel finds that policy guidance from Corps leadership is now quite clear regarding the need to pursue planning and regulatory activities in an integrated fashion, and that many on the Corps' planning and technical staff are motivated to carry out this mandate. At the same time, full and consistent implementation of this approach takes time and faces a variety of barriers. To provide a clear general direction, more complete and balanced guidance is needed regarding integrated planning, particularly in the evaluation of environmental, social, and economic costs and benefits and the identification of appropriate spatial and temporal scales at which different aspects of water projects must be evaluated. The Corps has identified serious and growing gaps in knowledge and capabilities and is taking steps to expand or reinstate training opportunities for its staff. These steps are essential and can help improve staff recruiting and retention, but more widespread use of outside expertise is also needed to complement in-house analytical capabilities in this increasingly complex planning environment. Beyond building analytical capabilities and lowering barriers to collaboration with other agencies, stakeholders, and outside experts, the most important step that can be taken to facilitate widespread implementation of integrated water resources planning is to promulgate clear standards requiring such an approach. If a clearly articulated and externally reviewed study plan of the salient social, economic, and environmental factors, at the appropriate spatial and temporal scales, is required to proceed to a feasibility study, and if a monitoring plan identifying key system properties that indicate project success is required to proceed to project implementation, project managers will have a clearer opportunity to implement existing Corps guidelines in the planning and evaluation of individual projects.

References

Allen, D.R., and M.N. Mayrega. 1970. The mechanics of compaction and rebound: Wilmington Oil Field, Long Beach, California. Pp. 410-423 in *Land Subsidence: Proceedings of the Tokyo Symposium*, September, 1969, Volume II. United National Educations, Scientific and Cultural Organization, Paris.

Arnold, E.J. 2002. State of the Lower Mississippi River Valley. Statement of Brigadier General E.J. Arnold to the Mississippi River Commission at Public Hearings, March 2002. Available [on-line] at *http://www.mvd.usace.army.mil /MRC/StateofValley/Spring2002.pdf* [accessed February 29, 2003].

Barras, J., S. Beville, D. Britsch, S. Hartley, S. Hawes, J. Johnson, P. Kemp, Q. Kinler, A. Martucci, J. Porthouse, D. Reed, K. Roy, S. Sapkota, and J. Suhayda. 2003. *Historical and Projected Coastal Louisiana Land Changes: 1978-2050*. USGS Open File Report 03-334. U.S. Geological Survey, Washington, D.C., 39 pp.

Bratkovich, A., S.P. Dinnel, and D.A. Goolsby. 1994. Variability and prediction of freshwater and nitrate fluxes for the Louisiana-Texas shelf. *Estuaries* 17(4):766-778.

Breitburg, D.L., and G.F. Riedel. In press. Multiple stressors in marine systems. In: *Marine Conservation Biology: The Science of Maintaining the Sea's Biodiversity*, E. Norse and L. Crowder, eds. Island Press, Washington, D.C.

Breitburg, D.L., J.G. Sanders, C.G. Gilmour, C.A. Hatfield, R.W. Osman, G.F. Riedel, S.P. Seitzinger, and K.G. Sellner. 1999. Variability in responses to nutrients and trace elements, and transmission of stressor effects through an estuarine food web. *Limnology and Oceanography* 44:837-863.

Brekke, L. 2002. Climatic Teleconnections for Influencing California Reservoir Operations: Identification, Assessment, and Constraints. Ph.D. dissertation, University of California, Berkeley.

Bricker, S.B., C.G. Clement, D.E. Pirhalla, S.P. Orlando, and D.R.G. Farrow. 1999. *National Estuarine Eutrophication Assessment: Effects of Nutrient Enrichment in the Nation's Estuaries*. National Oceanic and Atmospheric Administration, National Ocean Service, Special Projects Office and the National Centers for Coastal Ocean Science, Silver Spring, Maryland.

Bridge, J.S. 2003. *Rivers and Floodplains: Forms, Processes, and Sedimentary Record.* Blackwell Publishing, Oxford, United Kingdom.

Bruun, P. 1962. Sea level rise as a cause of shore erosion. *Journal of Waterways and Harbors Division* 88:117-130.

Burkett, V. 2001. Intertidal zones. The Earth system: Biological and ecological dimensions of global environmental change. Pp. 365-369 in *Encyclopedia of Global Environmental Change*, Volume 2, H.A. Mooney and J.G. Canadell, eds. John Wiley and Sons, Ltd., Chichester, United Kingdom.

Burkett, V.R., D.B. Zilkoski, and D.A. Hart. 2003. Sea-level rise and subsidence: Implications for flooding in New Orleans. Pp. 63-70 in *U.S. Geological Survey Subsidence Interest Group Conference, Proceedings of the Technical Meeting*, Galveston, Texas, November 27-29, 2001, K.R. Prince and D. L. Galloway, eds. Open File Report 03-308. U.S. Geological Survey, Water Resources Division, Washington, D.C.

California Reclamation Board and U.S. Army Corps of Engineers. 2002. The Sacramento and San Joaquin River Basin, California, Comprehensive Study, Interim Report. California Reclamation Board and U.S. Army Corps of Engineers, Sacramento, California.

Cayan, D.R., M.D. Dettinger, H.F. Diaz, and N.E. Graham. 1998. Decadal variability of precipitation over western North America. *Journal of Climate* 11:3148-3165.

Cayan, D.R., K.T. Redmond, and L.G. Riddle. 1999. ENSO and hydrologic extremes in the western United States. *Journal of Climate* 12:2881-2893.

Chesapeake Bay Program. 2000. Chesapeake 2000: A Watershed Partnership Agreement. U.S. Environmental Protection Agency, Annapolis, MD. Available [on-line] at *http://www.chesapeakebay.net/agreement.htm* [accessed March 17, 2004].

Clarke, J.N., and D. McCool. 1996. Staking Out the Terrain: Power and Performance in Natural Resource Agencies, 2nd ed. State University of New York Press, Albany, New York.

Coalition to Restore Coastal Louisiana. 1999. *No Time to Loose*. Baton Rouge, Louisiana, 66 pp.

Coastal Wetlands Conservation and Restoration Task Force. 1998. *Coast 2050: Toward a Sustainable Coastal Louisiana: An Executive Summary*. Louisiana Department of Natural Resources, Baton Rouge, 12 pp.

Committee on the Future of Coastal Louisiana. 2002. *Saving Coastal Louisiana: Recommendations for Implementing an Expanded Coastal Restoration Program*. State of Louisiana, Governor's Office of Coastal Activities, Baton Rouge, 47 pp.

Costanza, R., S.C. Farber, and J. Maxwell. 1989. The valuation and management of wetland ecosystems. *Ecological Economics* 1:335-361.

Culliton, T.J. 1998. Population: distribution, density and growth. NOAA State of the Coast Report. National Oceanic and Atmospheric Administration, Silver Spring, Maryland. Available [on-line] at *http://state-of-coast.noaa.gov/topics/html/pressure.html* [accessed March 25, 2004].

Day, J.H. 1981. The nature, origin and classification of estuaries. Chapter 1 in *Estuarine Ecology with Particular Reference to South Africa*, J.H. Day, ed. A.A. Balkema, Rotterdam.

Dean, R.G. 1995. Historical Shoreline Changes in the Vicinity of the P.E.P. Reef Installation and Reef Effects, Palm Beach, Florida. UFL/COEL-95/008. Coastal and Oceanographic Engineering Department, University of Florida, Gainesville.

Dean, R.G., and M. Perlin. 1977. Coastal engineering study of Ocean City Inlet, Maryland. Pp. 520-542 in Proceedings of the ASCE Specialty Conference on Coastal Sediments '77. American Society of Civil Engineers, Reston, Virginia.

Dean, R.G., O.H. Pilkey, Jr., and J.R. Houston. 1988. Eroding shorelines impose costly choices. *Geotimes* 33(5):9-14.

Del Rossi, A.F., and R.P. Inman. 1999. Changing the price of pork: The impact of local cost-sharing on legislator's demand for distributive public goods. *Journal of Public Economics* 71:247-73.

Dortch, Q., N.N. Rabalais, R.E. Turner, and G.T. Rowe. 1994. Respiration rates and hypoxia on the Louisiana shelf. *Estuaries* 17(4):862-872.

Ferguson, R.L. 2000. Proposal to change hydrologic unit delineations over the estuarine waters of North Carolina. White Paper. Available [on-line] at *http://www.ftw.nrcs.usda.gov/HUC/HU_Coastal_Proposal.pdf* [accessed February 20, 2004].

Goddard, L., S.J. Mason, S.E. Zebiak, C.F. Ropelewski, R. Basher, and M.A. Cane. 2001. Current approaches to seasonal to interannual climate predictions. *International Journal of Climatology* 21:1111-1152.

Griffin, Robert. 2002. Plenary Address. *Economic and Environmental Analysis Conference 2002: Balancing Economy and Environment*, New Orleans, Louisiana, July 16-18.

Hamlet, A.F., and D.P. Lettenmaier. 1999. Columbia River streamflow forecasting based on ENSO and PDO climate signals. *ASCE Journal of. Water Resources Planning and Management* 125:331-162.

Hansen, S., and C. Fischenich. 2002. An Assessment of Watershed Planning in Corps of Engineers Civil Works Projects, EMRRP Technical Notes Collection (ERDC TN-EMRRP-SR-34). U.S. Army Engineer Research and Development Center, Vicksburg, Mississippi.

Headland, J., P. Kotulak, A. Alcorn, J. Lesnick, P. McNeal, and S. Alfagame. 2000. Case histories for significant U. S. dredging projects. Chapter 10 in *Handbook of Dredging Engineering*, 2nd Ed., J.B. Herbich, ed. McGraw-Hill, New York.

Hicks, S.D., H.A. Debaugh, Jr., and L.E. Hickman. 1983. *Sea Level Variations for the United States, 1855-1980*. National Ocean Survey, Rockville, Maryland.

Hirsch, R.M., and J.M. Norris, 2001. *National Streamflow Information Program*. U.S.G.S. Fact Sheet FS-048-01. U.S. Geological Survey, Washington, D.C.

Hufschmidt, M.M. 1966. The Harvard program—A summing up. In *Water Research*, A.V. Kneese and S.C. Smith, eds. The Johns Hopkins University Press, Baltimore, Maryland.

Institute for Water Resources. 2002. *Trade-off Analysis and Procedures Guidebook*. IWR-02-R2. U.S. Army Institute for Water Resources, Decision Methodologies Division, Alexandria, Virginia.

Intergovernmental Panel on Climate Change (IPCC). 1995. *Climate Change 1995: Impacts, Adaptations and Mitigation of Climate Change: Scientific-Technical Analysis*. Cambridge University Press, Cambridge, Massachusetts, 878 pp.

Intergovernmental Panel on Climate Change (IPCC). 2001. *Climate Change 2001: The Scientific Basis. Contribution of Working Group 1 of the Third Assessment Report of the Intergovernmental Panel on Climate Change*, J.T. Houghton., Y. Ding, D.J. Griggs, M. Noguer, P.J. van der Linden, X. Dai, K. Maskell, and C.A. Johnson, eds. Cambridge University Press, Cambridge, United Kingdom, 881 pp.

Justic, D., N.N. Rabalais, R.E. Turner, and W.J. Wiseman. 1993. Seasonal coupling between riverborne nutrients, net productivity and hypoxia. *Marine Pollution Bulletin* 26:184-189.

Kahya, E., and J.A. Dracup. 1993. U.S. streamlfow patterns in relation to El Nino-Southern Oscillation. *Water Resources Research* 29:2491-2503.

Karl, T.R., and R.W. Knight. 1998. Secular trends of precipitation amount, frequency, and intensity in the United States. *Bulletin of the American Meteorological Society* 79(2):231-241.

Keeland, B.D., J.A. Allen, and V.R. Burkett. 1995. Southern forested wetlands. Pp. 216-218 in *Our Living Resources: A Report to the Nation on the Distribution, Abundance and Health of U. S. Plants, Animals and Ecosystems*, E.T. LaRoe, G.S. Farris, and P.T. Doran, eds. U.S. Department of Interior, National Biological Service, Washington, D.C.

Kenney, D.S. 1997. Resource management at the watershed level: An assessment of the changing federal role in the emerging era of community-based watershed management: Report to the Western Water Policy Review Advisory. UW 7.2000 R47. University of Colorado Law School, Natural Resources Law Center, Boulder, 15 pp.

Kennish, M.J. 1986. *Ecology of Estuaries*. CRC Press, Boca Raton, Florida, 254pp.

Loucks, D.P. 2003. Managing America's rivers: Who's doing it? *International Journal River Basin Management* 1:21-31.

Louisiana Department of Natural Resources. 2002. Looking to the future. *Water Marks* 21:21.

Maass, A., M.M. Hufschmidt, R. Dorfman, H.A. Thomas Jr., S.A. Marglin, and G.M. Fair. 1962. *Design of Water-Resource Systems*. Harvard University Press, Cambridge, Massachusetts.

Mann, R. 2000. Restoring the oyster reef communities in the Chesapeake Bay: A commentary. *Journal of Shellfish Research* 19:335-339.

Martin, L.R. 2002. *Regional Sediment Management: Background and Overview of Initial Implementation*. IWR Report 02-PS-2. U.S. Army Corps of Engineers, Institute for Water Resources, Alexandria, Virginia.

McDonald, P.O., W.E. Frayer, and J.K. Klausen. 1979. *Documentation, Chronology, and Future Projections of Bottomland Hardwood Habitat Losses in the Lower Mississippi River Alluvial Plain*, 2 volumes, U.S. Department of Interior, Fish and Wildlife Service, Vicksburg, Mississippi.

Meade, R.H. 1995. Geology, hydrology, sediments and engineering of the Mississippi River. Contaminants in the Mississippi River, 1987-1992, R.H. Meade, ed. *U.S. Geological Survey Circular* 1133:3-31.

Melillo, J.M., A.C. Janetos, T.R. Karl, R.C. Corell, E.J. Barron, V. Burkett, T.F. Cecich, K. Jacobs, L. Joyce, B. Miller, M.G. Morgan, E.A. Parson, R.G. Richels, and D.S. Schimel. 2000. *Climate Change Impacts on the United States: The Potential Consequences of Climate Variability and Change, Overview*. Cambridge University Press, Cambridge, United Kingdom, 154 pp.

Miller, D.C., S.L. Poucher, and L. Coiro. 2002. Determination of lethal dissolved oxygen levels for selected marine and estuarine fishes, crustaceans, and a bivalve. *Marine Biology* 140:287-296.

Mitsch, W.J., and S.E. Jørgensen. 2003. Ecological engineering: A field whose time has come. *Ecological Engineering* 20:363-377.

Mitsch, W.J., and S.E. Jørgensen. 2004. *Ecological Engineering and Ecosystem Restoration*. John Wiley & Sons, Inc., New York, 411 pp.

Muir, John. 1911. My First Summer in the Sierra. Houghton Mifflin Company, New York, 400 pp.

National Marine Fisheries Service (NMFS). 1998. *Ecosystem-Based Fishery Management*. A Report to Congress by the Ecosystem Principles Advisory Panel. National Marine Fisheries Service, Silver Spring, Maryland, 53 pp.

National Research Council (NRC). 1992. *Restoration of Aquatic Ecosystems*. National Academy Press, Washington, D.C.

National Research Council (NRC). 1996. *Upstream: Salmon and Society in the Pacific Northwest*. National Academy Press, Washington, D.C., 452 pp.

National Research Council (NRC). 1999a. *New Strategies for America's Watersheds*. National Academy Press. Washington, D.C., 328 pp.

National Research Council (NRC). 1999b. *New Directions in Water Resources Planning for the U.S. Army Corps of Engineers*. National Academy Press, Washington, D.C., 120 pp.

National Research Council (NRC). 2000. *Clean Coastal Waters: Understanding and Reducing the Effects of Nutrient Pollution*. National Academy Press, Washington, D.C., 428 pp.

National Research Council (NRC). 2001a. *Compensating for Wetland Losses Under the Clean Water Act*. National Academy Press. Washington, D.C., 322 pp.

National Research Council (NRC). 2001b. *Climate Change Science: An Analysis of Some Key Questions*. National Academy Press, Washington, D.C. , 29 pp.

National Research Council (NRC). 2002a. *Review Procedures for Water Resources Project Planning*. The National Academies Press, Washington, D.C.

National Research Council (NRC). 2002b. *The Missouri River Ecosystem: Exploring the Prospects for Recovery*, The National Academies Press, Washington, D.C.

National Research Council (NRC). 2004a. *Adaptive Management for Water Resources Project Planning*. The National Academies Press, Washington, D.C.

National Research Council (NRC). 2004b. *Analytical Methods and Approaches for Water Resources Project Planning*. The National Academies Press, Washington, D.C.

National Research Council (NRC) 2004c. *A Geospatial Framework for the Coastal Zone: National Needs for Coastal Mapping and Charting*. The National Academies Press, Washington, D.C.

Nichols, M.D. 2003. Draft Review of California Coastal Erosion Planning and Response: A Strategy for Action. California Resources Agency, Sacramento.

Onuf, C.P. 1994. Seagrasses, dredging, and light in Laguna Madre, Texas. *Coastal and Shelf Science* 39:75-91.

Piechota, T.C., and J.A. Dracup. 1996. Drought and regional hydrologic variations in the United States: Associations with the El Nino/Southern Oscillation. *Water Resources Research* 32(5):1359-1373.

Quammen, M.L., and C.P. Onuf. 1993. Laguna Madre: Seagrass changes continues decades after salinity reduction. *Estuaries* 16:302-310.

Rabalais, N.N., and R.E.Turner. 2001. Hypoxia in the Northern Gulf of Mexico: Description, causes and change. Pp. 1-36 in *Coastal Hypoxia: Consequences for Living Resources and Ecosystems*. American Geophysical Union, Washington, D.C.

Rabalais, N.N., R.E. Turner, and D. Scavia. 2002. Beyond science into policy: Gulf of Mexico hypoxia and the Mississippi River. *BioScience* 52:129-142.

Raffaelli, D., I. Karakassis, and A. Galloway. 1991. Zonation schemes on sandy shores: A multivariate approach. *Journal of Experimental Marine Biology and Ecology* 148:241-253.

Rothschild, B.J., J.S. Ault, P. Goulletquer, and M. Heral. 1994. Decline of the Chesapeake Bay oyster population: A century of habitat destruction and overfishing. *Marine Ecology Progress Series* 111:29-39.

Schad, T.M. 1998. Water policy: Who should do what? Water Resources Update. Issue 111. Universities Council on Water Resources, Carbondale, Illinois.

Seaber, P.R., F.P. Kapinos, and G.L. Knapp. 1987. *Hydrologic Units Maps*. U.S.G.S. Water-Supply Paper 2294. U.S. Geological Survey, Washington, D.C., 63 pp.

Shabman, L., and L. Zepp. 2000. An Approach for Evaluating Nonstructural Actions with Application to the Yazoo River (Mississippi). A report prepared in cooperation with the U.S. Environmental Protection Agency, Region 4, Atlanta, Georgia, 140 pp.

Stachowicz, J.J., J.R. Terwin, R.B. Whitlatch and R.W. Osman. 2003. Linking climate change and biological invasions: Ocean warming facilitates nonindigenous species invasions. Available [on-line] at *http://www.acnatsci.org/press/special/OsmanERC.pdf* [accessed March 24, 2004].

Stakhiv, E.Z. 2003. Disintegrated water resources management in the U.S.: Union of Sisyphus and Pandora. *Journal of Water Resources Planning and Management* 129(3):151-154 (H 31598).

Texas Department of Transportation. 2000. *The Gulf Intracoastal Waterway in Texas*. Texas Department of Transportation, Austin, 109 pp.

Thompson, S. 1995. Methods of environmental impact assessment. Pp. 278-296 in *The Natural and Built Environment Series*, P. Morris and R. Therivel, eds. University College London Press, Ltd., United Kingdom.

Tiner, R.W. 1984. *Wetlands of the United States: Current Status and Future Trends*. U.S. Department of Interior, Fish and Wildlife Service, Newton Corner, Massachusetts, 59 pp.

Turner, R.E., A.M. Redmond, and J.B. Zedler. 2001. Count it by acre or function: Mitigation adds up to net loss of wetlands. *National Wetlands Newsletter* 23: 5-6, 14-16.

U.S. Army Corps of Engineers. 1959. Vicksburg District, Mississippi River and Tributaries Comprehensive Review Report, Annex L, Yazoo Backwater Project Mississippi, November 20.

U.S. Army Corps of Engineers. 1979. *Missouri River Main Stem Reservoir System Reservoir Regulation Manual* (Master Manual). U.S. Army Engineer Division, Missouri River Corps of Engineers, Omaha, Nebraska.

U.S. Army Corps of Engineers. 1997. USACE, IWR, Planning Primer. November. IWR Report 97-R-15. U.S. Army Corps of Engineers, Institute for Water Resources, Alexandria, Virginia, 13 pp. Available [on-line] at *http://www.iwr.usace.army.mil/iwr/products/reports/reports.htm* [accessed March 24, 2004].

U.S. Army Corps of Engineers. 1999a. Ecosystem Restoration—Supporting Policy Information. EP-1165-2-502, 1999. Available [on-line] at *http://www.usace army.mil/inet/usace-docs/eng-pamphlets/ep1165-2-502/toc.htm* [accessed April 1, 2004].

U.S. Army Corps of Engineers. 1999b. Digest of Water Resources Policies and Authorities. 30 July 1999. EP 1165-2-1. Available [on-line] at *http://www.usace.army.mil/inet/usace-docs/eng-pamphlets/ep1165-2-1/toc.htm* [accessed March 24, 3004].

U.S. Army Corps of Engineers. 1999c. Habitat Impacts of the Construction of the MRGO, prepared for the Environmental Subcommittee of the Technical Committee convened by the Environmental Protection Agency in response to the St. Bernard Parish Council Resolution 12-98. U.S. Army Corps of Engineers, New Orleans District.

U.S. Army Corps of Engineers. 1999d. Policy Guidance Letter 61: Application of Watershed Perspective to Corps of Engineers Civil Works Programs and Activities. CECW-AA. Available [on-line] at *http://www.usace.army.mil.inet/functions/cw/cecwp/branches/guidance_dev/pgls/pdf/pgl61.pdf* [accessed April 4, 2004].

U.S. Army Corps of Engineers. 2000a. *Draft Yazoo Backwater Area Reformulation Report*, 3 vol. U.S. Army Corps of Engineers, Vicksburg, Mississippi.

U.S. Army Corps of Engineers. 2000b. *Planning Guidance Notebook.* 22 April 2000. Publication Number ER 1105-2-100. U.S. Army Corps of Engineers, Washington, D.C.

U.S. Army Corps of Engineers. 2000c. *Planning and Guidance Notebook.* [abridged version of 2000b].

U.S. Army Corps of Engineers. 2001a. Environmental Operating Principles. Available [on-line] at *http://www.hq.usace.army.mil/cepa/envprinciples.htm* [accessed February 20, 2004].

U.S. Army Corps of Engineers. 2001b. Project Partnership Kit. IWR Report No. 96-R-10. U.S Army Corps of Engineers, Institute of Water Resources, Alexandria, Virginia. Available [on-line] at *http://www.iwr.usace.army.mil/iwr/pdf/ppkit.pdf* [accessed March 16, 2004].

U.S. Army Corps of Engineers. 2002a. Gee whiz—Facts and figures about the Mississippi Valley Division. Mississippi Valley Division, Vicksburg, 2 pp.

U.S. Army Corps of Engineers. 2002b. Program Management Plan (PgMP) for Integrating the USACE Environmental Operating Principles (EOPs). U.S. Army Corps of Engineers, Jacksonville District, South Atlantic Division.

U.S. Army Corps of Engineers. 2002c. September 2002 draft. Department of the Army Corps of Engineers Civil Works Program Strategic Plan FY 2003-FY2008.

U.S. Army Corps of Engineers. 2002d. Fact Sheet. U.S. Army Corps of Engineers, Regional Sediment Management Research Program. Available [on-line] at *http://chl.wes.army.mil/research/sedimentation/RSM/FactSheetRSMP.pdf* [accessed April 1, 2004].

U.S. Army Corps of Engineers. 2002e. Baltimore District—Poplar Island Environmental Restoration, Project Brief, Baltimore District. Available [on-line] at *http://www.nab.usace.army.mil/projects/Maryland/poplarisland.htm* [accessed April 1, 2004].

U.S. Army Corps of Engineers. 2002f. Program Description (Draft), RSM Draft Technical Report 1 January 2002. U.S. Army Corps of Engineers, Regional Sediment Management Research Program. Available [on-line] at *http://www.wes.army.mil/rsm/pubs/reports.html* [accessed February 20, 2004].

U.S. Army Corps of Engineers. 2002g. U.S. Army Corps of Engineers Operating Principles and Implementation Guidance. March 26.

U.S. Army Corps of Engineers. 2003a. U.S. Army Corps of Engineers Regulatory Program: All Permit Decisions FY 2002. Available [on-line] at *http://www.usace.army.mil/inet/functions/cw/cecwo/reg/2002webcharts.pdf* [accessed February 20, 2004].

U.S. Army Corps of Engineers. 2003b. Mississippi River and Tributaries Project. New Orleans District. Brochure.

U.S. Army Corps of Engineers. 2003c. *Program Management Plan for Integrating the Environmental Operating Principles Within HQUSACE.* Final Draft. U.S. Army Corps of Engineers, Washington, D.C.

U.S. Department of Agriculture. 2003a. Federal Standards for Delineation of Hydrologic Unit Boundaries—March 1, 2002-March 1, 2003. U.S. Department of Agriculture, Natural Resources Conservation Service, Federal Geographic Data Committee. Available [on-line] at *http://www.ftw.nrcs.usda.gov/huc_data.html* [accessed May 23, 2003].

U.S. Department of Agriculture. 2003b. Mapping and Digitizing Watershed and Subwatershed Hydrologic Unit Boundaries. U.S. Department of Agriculture, Natural Resources Conservation Service, Washington, D.C. Available [on-line] at *http://www.ftw.nrcs.usda.gov/huc/ni170304.html* [accessed April 1, 2004].

U.S. Department of the Army. 2001. Fiscal Year 2002, Civil works budget for the U.S. Army Corps of Engineers. U.S. Department of the Army, Office of the Assistant Secretary of the Army (Civil Works), Washington, D.C.

U.S. Department of the Army. 2003. Fiscal Year 2004, Civil works budget for the U.S. Army Corps of Engineers. U.S. Department of the Army, Office of the Assistant Secretary of the Army (Civil Works), Washington, D.C.

U.S. Environmental Protection Agency (USEPA). 1999. *The Ecological Condition of Estuaries in the Gulf of Mexico*. EPA620-R-98-004, Environmental Projection Agency, Washington, D.C., 71 pp.

U.S. Environmental Protection Agency (USEPA). 2000. Yazoo Backwater Area: Technical Review of the Draft Reformulation Report. U.S. Environmental Protection Agency, Washington, D.C.

U.S. Fish and Wildlife Service (USFWS). 2000a. Letter from the U.S. Fish and Wildlife Service, Jackson, Mississippi, to the U.S. Army Corps of Engineers, Vicksburg District, dated October 16.

U.S. Fish and Wildlife Service (USFWS). 2000b. Letter with attachments from the U.S. Department of Interior, Office of the Secretary, Office of Environmental Policy and Compliance, Atlanta, Georgia, to the U.S Army Corps of Engineers, District Engineer, Vicksburg District, dated November 3.

U.S. Fish and Wildlife Service (USFWS). 2000c. Letter with attachments from the U.S. Fish and Wildlife Service, Jackson, Mississippi, to the U.S Army Corps of Engineers, District Engineer, Vicksburg District, dated October 16.

U.S Fish and Wildlife Service (USFWS). 2003. Perspective on the Corps of Engineers Proposed Yazoo Pumps Project. U.S. Fish and Wildlife Service, Jackson Ecological Services Field Office, Jackson, Mississippi.

U.S. Senate, Committee on Environment and Public Works. 2002. Water Resources Development Act Issues. Hearing before the Committee on Environment and Public Works. Testimony of Robert B. Flowers on the Water Resources Development Programs within the U.S. Army Corps of Engineers, June 18, 2002, 107th Congress, 2nd Session. U.S. Government Printing Office, Washington, D.C. Available [on-line] at *http://www.senate.gov/~epw/107th/Flowers_061802.htm* [accessed February 17, 2004].

Walker, J.R., and J.W. Dunham. 1977. Lake Worth Inlet—Case study. Pp. 602-621 in *Proceedings of Coastal Sediments '77*. American Society of Civil Engineers, Charleston, S.C.

Watzin, M.C., and J.G. Gosselink. 1992. The Fragile Fringe: Coastal Wetlands of the United States. Louisiana Sea Grant College Program, Louisiana State University, Baton Rouge., 16 pp.

Water Resources Council (WRC). 1983. *Economic and Environmental Principles and Guidelines for Water and Related Land Resources Implementation Studies*. U.S. Government Printing Office, Washington, D.C.

Weinstein, M.P., J.H. Balletto, J.M. Teal, and D.F. Ludwig. 1997. Success criteria and adaptive management for a large-scale wetland restoration project. *Wetlands Ecology and Management* 4:111-127.

Weinstein, M.P., J.M. Teal, J.H. Balletto, and K.A. Strait. 2001. Restoration principles emerging from one of the world's largest tidal marsh restoration projects. *Wetlands Ecology and Management* 9:387-407.

Appendixes

Appendix A

Panel and Staff Biographies

PANEL MEMBERS

Peter R. Wilcock, *Chair*, is a professor in the Department of Geography and Environmental Engineering at The Johns Hopkins University. He received his B.S. in physical geography from the University of Illinois, Urbana-Champaign; his M.S. in geomorphology from McGill University; and his Ph.D. in geology from the Massachusetts Institute of Technology. His research focuses on problems of erosion and sedimentation and their application to river management. In 1997-1998, Dr. Wilcock served on the National Research Council (NRC) committee evaluating adaptive management of Glen Canyon Dam on the Colorado River ecosystem.

Gail M. Ashley is a professor of geological sciences at Rutgers University, New Brunswick, New Jersey. She is also director of the Quaternary Studies Graduate Program. Her research interests include a comparison of terrestrial records of paleoclimate during the Quaternary in polar, temperate, and tropical regions and reconstruction of the paleoenvironment of early hominids. She has served as president of the Geological Society of America, vice-president of the International Association of Sedimentologists and editor-in-chief of the *Journal of Sedimentary Research*, president of the Society for Sedimentary Geology, and a past member of the Board of Earth Sciences and Resources.

Denise L. Breitburg is a senior scientist at the Smithsonian Environmental Research Center and an adjunct professor at the University of Maryland. Formerly, she worked as a curator at the Academy of Natural Sciences, Estuarine Research Center. Her research integrates aspects of ecology, animal behavior, and the effects of human population on marine and estuarine systems. She received a B.S. in biology from Arizona State University, and an M.A. and Ph.D. in biology from the University of California, Santa Barbara.

Virginia R. Burkett is chief of the Forest Ecology Branch of the National Wetlands Research Center at Lafayette, Louisiana. Dr. Burkett's current research involves climate change impacts in coastal regions and bottomland hardwood regeneration in frequently flooded sites of the Mississippi River alluvial valley. Before joining the U.S. Geological Survey, she was director of the Louisiana Department of Wildlife and Fisheries. She has also served as assistant director of the Louisiana Geological Survey and as director of the Louisiana Coastal Zone Management Program. She received her B.S. and M.S. from Northwestern State University of Louisiana and her doctorate degree in forestry from Stephen F. Austin State University in Nacogdoches, Texas.

Joseph L. Cordes is professor of economics at George Washington University. He directs the Ph.D. program in public policy. Dr. Cordes served as associate dean for faculty affairs and programs in the Columbian School of Arts and Sciences and as chair of the Department of Economics from 1991 to 1997. He has also held appointments as an economic policy fellow at the Brookings Institute, a financial economist at the U.S. Department of the Treasury, and deputy assistant director for tax analysis at the Congressional Budget Office. His research focuses on the economic behavior of nonprofit organizations and on evaluating public programs intended to reduce economic risks from flood and storm damage. He received an M.S. and a Ph.D. in economics from the University of Wisconsin, Madison.

Robert G. Dean has been graduate research professor of coastal and ocean engineering at the University of Florida since 1982. Previously, he held faculty positions at the University of Delaware, the University of Washington, and the Massachusetts Institute of Technology (MIT). He also has served as a consultant on coastal and ocean engineering to private industry and government. Dr. Dean is an expert in wave mechanics and coastal engineering problems, and he has published many papers on wave theory, beach erosion problems, tidal inlets, and coastal structures. He is a past recipient of the John G. Moffatt-Frank E. Nichol Harbor and Coastal Engineering Award given by the American Society of Civil Engineers. He is a member of the National Academy of Engineering and chair of the NRC Committee on the Restoration and Protection of Coastal Louisiana. Dr. Dean is also a former member of the National Academies' Marine Board and a former chair of the NRC Committee on Engineering Implications of Sea Level Rise. Dr. Dean has a B.S. from the University of California, Berkeley; an M.S. from the Agricultural and Mechanical College of Texas; and a Sc.D. in civil engineering from MIT.

John A. Dracup is a professor in the department of civil and environmental engineering at the University of California, Berkeley. Previously, he served on the faculty of the University of California, Los Angeles. His research interests include hydroclimatology; analysis of large-scale water resource systems and hydrologic and environmental systems; engineering economics of water resources systems; and surface water hydrology. He served as lieutenant with the U.S. Army Corps of Engineers from 1957 to 1958. He received a B.S. from the University of Washington, an M.S. from MIT, and a Ph.D. from the University of California, Berkeley.

William J. Mitsch is a professor of natural resources and environmental science at Ohio State University and director of the Olentangy River Wetland Research Park. His research interests include wetland ecology and management; wetland restoration and creation; wetland biogeochemistry; ecological economics of wetlands and other ecosystems; ecological engineering; ecosystem ecology and modeling; wetlands and global climate change; wetland vegetation dynamics; and primary productivity in aquatic systems. Dr. Mitsch is editor-in-chief of *Ecological Engineering*. He served as a member of the NRC's committees on characterization of wetlands and mitigating wetland losses. He received his Ph.D. in environmental engineering sciences (systems ecology) from the University of Florida.

Robert E. Randall is professor of ocean and civil engineering at Texas A&M University, where he directs the university's Center for Dredging Studies. He has served at head of the Ocean Engineering Program and Ocean and Hydraulic Engineering Group in the Civil Engineering Department. He was an ocean engineer at Harbor Branch Foundation, Inc., from 1973 to 1975 and a mechanical engineer at the Naval Underwater System Center from 1972 to 1973. Dr. Randall is a member of the American Society of Civil Engineers, Society of Naval Architects and

Marine Engineers, Western Dredging Association, American Society of Engineering Education, American Society of Mechanical Engineers, and a fellow in the Marine Technology Society. He received his B.M.E. in mechanical engineering from Ohio State University and his M.S. and Ph.D. in ocean engineering from the University of Rhode Island.

A. Dan Tarlock holds an A.B. and LL.B. from Stanford University and is currently distinguished professor of law at the Chicago-Kent College of Law. He previously practiced law in San Francisco and Denver, and taught at the University of Chicago, University of Hawaii, Indiana University, University of Kansas, University of Michigan, University of Texas, and University of Utah. Mr. Tarlock has written and consulted widely in the fields of water law, environmental protection, and natural resources management. Mr. Tarlock served as a member of the Water Science and Technology Board and chaired the Committee on Western Water Management Change, which published the report *Water Transfers in the West*. In 1997 to 1998, he served as the principal writer for the Western Policy Advisory Review Commission's report *Water in the West*. Mr. Tarlock is currently serving as one of the three United States legal advisers to the Secretariat of the Commission on Environmental Cooperation, established by the North American Free Trade Agreement Environmental Side Agreement.

NRC STAFF MEMBERS

John Dandelski (*study director*) received his M.A. in marine affairs and policy from the Rosenstiel School of Marine and Atmospheric Science (RSMAS), University of Miami, in 2001, where his research focused on evaluating fisheries' impacts to the benthic communities of Biscayne Bay. While at RSMAS he served as the assistant diving safety officer and also worked for the International Oceanographic Foundation. Mr. Dandelski has been a research associate with the Ocean Studies Board since 2001. As a graduate research intern at the Congressional Research Service he wrote two reports for Congress on fisheries and ocean health issues. He was also the project manager for the NRC *Abrupt Climate Change: Inevitable Surprises* and has worked on a number of other NRC reports, including *Environmental Information for Naval Warfare* (2003), and *Oil in the Sea III: Inputs, Fates, and Effects* (2003). Mr. Dandelski also holds an M.S. in industrial/organizational psychology and worked in the field of experiential environmental education for several years.

Dan Walker (*senior program officer*) obtained his Ph.D. in geology from the University of Tennessee in 1990. A senior program officer at the Ocean Studies Board, Dr. Walker also holds a joint appointment as a guest investigator at the Marine Policy Center of the Woods Hole Oceanographic Institution. Since joining the Ocean Studies Board in 1995, he has directed a number of studies including *Environmental Information for Naval Warfare* (2003); *Oil in the Sea III: Inputs, Fates and Effects* (2003); *Clean Coastal Waters: Understanding and Reducing the Effects of Nutrient Pollution* (2000); *Science for Decisionmaking: Coastal and Marine Geology at the U.S. Geological Survey* (1999); *Global Ocean Sciences: Toward an Integrated Approach* (1998); and *The Global Ocean Observing System: Users, Benefits, and Priorities* (1997). A member of the American Geophysical Union, the Geological Society of America, and the Oceanography Society, Dr. Walker was recently named editor of the *Marine Technology Society Journal*. A former member of both the Kentucky and the North Carolina State geologic surveys, Dr. Walker's interests focus on the value of environmental information for policy making at local, state, and national levels.

Nancy A. Caputo (*senior project assistant*) received a master's of public policy from the University of Southern California and a bachelor's degree in political science-international relations from the University of California, Santa Barbara. During her tenure with the Ocean Studies Board, she has assisted with the completion of four reports: *A Review of the Florida Keys Carrying Capacity Study* (2002); *Emulsified Fuels—Risks and Response* (2002); *Decline of the Steller Sea Lion in Alaskan Waters—Untangling Food Webs and Fishing Nets* (2003); and *Enabling Ocean Research in the 21st Century: Implementation of a Network of Ocean Observatories* (2003). Ms. Caputo has previous professional experience in researching fisheries management in the northeastern and northwestern United States, socioeconomic assistance programs for fishing communities, and habitat restoration programs. Her interests include marine policy and science, oceanographic education, coastal management, and habitat restoration.

Appendix B

Acronyms

AFB	alternative formulation briefing
BCA	benefit-cost analysis
CRP	Conservation Reserve Program
CSMW	Coastal Sediment Management Workgroup
CVP	Central Valley Project
CWA	Clean Water Act
CWPPRA	Coastal Wetlands Planning, Protection, and Restoration Act
EIS	environmental impact statement
ENSO	El Niño Southern Oscillation
FCA	Flood Control Act
FCSA	feasibility cost-sharing arrangement
FGDC	Federal Geographic Data Committee
FY	fiscal year
GCM	General Circulation Model
GIS	geographic information systems
GIWW	Gulf Intracoastal Waterway
HEC	Hydrologic Engineering Center
HUC	hydrologic unit code

IPCC	Intergovernmental Panel on Climate Change
ITR	independent technical review
IWR	Institute for Water Resources
LCA study	Louisiana Comprehensive Coastwide Ecosystem Restoration Study
LERRDS	lands, easements, rights of way, relocations, and disposal areas
MR&T	Mississippi River and Tributaries
MOA	memorandum of agreement
MOU	memorandum of understanding
MRGO	Mississippi River Gulf Outlet
MSL	mean sea level
NED	national economic development
NEPA	National Environmental Policy Act
NER	national ecosystem restoration
NGO	nongovernmental organization
NMFS	National Marine Fisheries Service
NRC	National Research Council
NRCS	Natural Resources Conservation Service
O&M	operation and maintenance
OMB	Office of Management and Budget
P&G	*Economic and Environmental Principles and Guidelines for Water and Related Land Resources Implementation Studies*
P&S	*Principles and Standards for Planning Water and Related Land Resources*
PED	pre-construction engineering and design
PGN	*Planning Guidance Notebook*
PNW	Pacific Northwest
PPE	post-project evaluation
PSE&G	Public Service Electric and Gas Company
PSP	project study plan
PSW	Pacific Southwest
RSL	relative sea level
RSMP	Regional Sediment Management Program

SWP State Water Project

TVA Tennessee Valley Authority

UMRBA Upper Mississippi River Basin Association
USEPA U.S. Environmental Protection Agency
USFWS U.S. Fish and Wildlife Service
USGS U.S. Geological Survey

WRC Water Resources Council
WRDA Water Resources Development Act
WRP Wetlands Reserve Program
WRPA Water Resources Planning Act

Appendix C

Rosters

Coordinating Committee
LEONARD SHABMAN, *Chair*, Resources for the Future, Washington, D.C.
GREGORY B. BAECHER, University of Maryland, College Park
DONALD F. BOESCH, University of Maryland, Cambridge
ROBERT W. HOWARTH, Cornell University, Ithaca, New York (*through November 2002*)
GERALDINE KNATZ, Port of Long Beach, Long Beach, California
JAMES K. MITCHELL, Virginia Polytechnic Institute and State University, Blacksburg
LARRY A. ROESNER, Colorado State University, Fort Collins (*through August 2003*)
A. DAN TARLOCK, Chicago-Kent College of Law, Chicago, Illinois
VICTORIA J. TSCHINKEL, The Nature Conservancy, Altamonte Springs, Florida
JAMES G. WENZEL, Marine Development Associates, Inc., Saratoga, California
M. GORDON WOLMAN, Johns Hopkins University, Baltimore, Maryland

Peer Review Procedures
JAMES K. MITCHELL, *Chair*, Virginia Polytechnic Institute and State University, Blacksburg
MELBOURNE BRISCOE, Office of Naval Research, Arlington, Virginia

STEPHEN J. BURGES, University of Washington, Seattle
LINDA CAPUANO, Honeywell, Inc., San Jose, California
DENISE FORT, University of New Mexico, Albuquerque
PORTER HOAGLAND, Woods Hole Oceanographic Institution, Massachusetts
DAVID H. MOREAU, University of North Carolina, Chapel Hill
CRAIG PHILIP, Ingram Barge Company, Nashville, Tennessee
JOHN T. RHETT, Consultant, Arlington, Virginia
RICHARD E. SPARKS, Illinois Water Resources Center, Urbana
BORY STEINBERG, Steinberg and Associates, McLean, Virginia

Panel on Methods and Techniques of Project Analysis
GREGORY B. BAECHER, *Chair*, University of Maryland, College Park
JOHN B. BRADEN, University of Illinois, Urbana-Champaign
DAVID L. GALAT, University of Missouri, Columbia
GERALD E. GALLOWAY, Titan Corporation, Fairfax, Virginia
ROBERT G. HEALY, Duke University, Durham, North Carolina
EDWIN E. HERRICKS, University of Illinois, Urbana-Champaign
CATHERINE L. KLING, Iowa State University, Ames
LINDA A. MALONE, College of William and Mary, Williamsburg, Virginia
RAM MOHAN, Blasland, Bouck & Lee, Inc., Annapolis, Maryland
MAX J. PFEFFER, Cornell University, Ithaca, New York
DOUG PLASENCIA, AMEC, Phoenix, Arizona
DENISE J. REED, University of New Orleans, Louisiana
JAN A. VELTROP, Consultant, Skokie, Illinois

Adaptive Management for Resource Stewardship
DONALD F. BOESCH, *Chair*, University of Maryland-Center for Environmental Science, Cambridge
HENRY J. BOKUNIEWICZ, University of New York, Stony Brook
RICHARD DE NEUFVILLE, Massachusetts Institute of Technology, Cambridge
G. EDWARD DICKEY, Consultant, Baltimore, Maryland
HOLLY D. DOREMUS, University of California, Davis
FREDRICK J. HITZHUSEN, Ohio State University, Columbus
CARL HERSHNER, Virginia Institute of Marine Science, Gloucester Point, Virginia
CHARLES D. D. HOWARD, Charles Howard Associates, British Columbia, Canada

WILLIAM R. LOWRY, Washington University, St. Louis, Missouri
BARRY NOON, Colorado State University, Fort Collins
THAYER SCUDDER, California Technology Institute, Pasadena
ROBERT W. STERNER, University of Minnesota, Minneapolis

River Basin and Coastal Systems Planning
PETER R. WILCOCK, *Chair*, Johns Hopkins University, Baltimore, Maryland
GAIL M. ASHLEY, Rutgers University, Piscataway, New Jersey
DENISE L. BREITBURG, Smithsonian Environmental Research Center, Edgewater, Maryland
VIRGINIA R. BURKETT, U.S. Geological Survey, Lafayette, Louisiana
JOSEPH J. CORDES, George Washington University, Washington, D.C.
ROBERT G. DEAN, University of Florida, Gainesville
JOHN A. DRACUP, University of California, Berkeley
WILLIAM J. MITSCH, Ohio State University, Columbus
ROBERT E. RANDALL, Texas A&M University, College Station
A. DAN TARLOCK, Chicago Kent College of Law, Chicago, Illinois

Water Science and Technology Board
RICHARD G. LUTHY, *Chair*, Stanford University, Stanford, California
JOAN B. ROSE, *Vice Chair*, Michigan State University, East Lansing
RICHELLE M. ALLEN-KING, State University of New York (SUNY), Buffalo, New York
GREGORY B. BAECHER, University of Maryland, College Park
KENNETH R. BRADBURY, Wisconsin Geological and Natural History Survey, Madison
JAMES CROOK, Water Reuse Consultant, Norwell, Massachusetts
EFI FOUFOULA-GEORGIOU, University of Minnesota, Minneapolis
PETER GLEICK, Pacific Institute for Studies in Development, Environment, and Security, Oakland, California
JOHN LETEY, JR., University of California, Riverside
DIANE M. MCKNIGHT, University of Colorado, Boulder (*through June 30, 2003*)
CHRISTINE L. MOE, Emory University, Atlanta, Georgia
ROBERT PERCIASEPE, National Audubon Society, Washington, D.C.
RUTHERFORD H. PLATT, University of Massachusetts, Amherst (*through June 30, 2003*)
JERALD L. SCHNOOR, University of Iowa, Iowa City
LEONARD SHABMAN, Resources for the Future, Washington, D.C.

R. RHODES TRUSSELL, Trussell Technologies, Inc., Pasadena, California
KARL K. TUREKIAN, Yale University, New Haven, Connecticut
HAME M. WATT, Independent Consultant, Washington, D.C.
JAMES L. WESCOAT, JR., University of Illinois at Urbana-Champaign

Water Science and Technology Board Staff
STEPHEN D. PARKER, Director
LAURA J. EHLERS, Senior Staff Officer
JEFFREY W. JACOBS, Senior Staff Officer
WILLIAM S. LOGAN, Senior Staff Officer
LAUREN E. ALEXANDER, Staff Officer
MARK C. GIBSON, Staff Officer
STEPHANIE E. JOHNSON, Staff Officer
M. JEANNE AQUILINO, Administrative Associate
ELLEN A. DE GUZMAN, Research Associate
PATRICIA JONES KERSHAW, Study/Research Associate
ANITA A. HALL, Administrative Assistant
JON Q. SANDERS, Senior Project Assistant
DOROTHY K. WEIR, Project Assistant

Ocean Studies Board
NANCY RABALAIS, *Chair*, Louisiana Universities Marine Consortium, Chauvin
LEE G. ANDERSON, University of Delaware, Newark
WHITLOW AU, University of Hawaii at Manoa
ARTHUR BAGGEROER, Massachusetts Institute of Technology, Cambridge
RICHARD B. DERISO, Inter-American Tropical Tuna Commission, La Jolla, California
ROBERT B. DITTON, Texas A&M University, College Station
EARL DOYLE, Shell Oil (Retired), Sugar Land, Texas
ROBERT DUCE, Texas A&M University, College Station
PAUL G. GAFFNEY, II, National Defense University, Washington, D.C.
WAYNE R. GEYER, Woods Hole Oceanographic Institution, Massachusetts
STANLEY R. HART, Woods Hole Oceanographic Institution, Massachusetts
RALPH S. LEWIS, Connecticut Geological Survey (Retired), Hartford
WILLIAM F. MARCUSON, III, U.S. Army Corp of Engineers

(Retired), Vicksburg, Mississippi
JULIAN P. MCCREARY, JR., University of Hawaii, Honolulu
JACQUELINE MICHEL, Research Planning, Inc., Columbia, South Carolina
JOAN OLTMAN-SHAY, Northwest Research Associates, Inc., Bellevue, Washington
ROBERT T. PAINE, University of Washington, Seattle
SHIRLEY A. POMPONI, Harbor Branch Oceanographic Institution, Fort Pierce, Florida
FRED N. SPIESS, Scripps Institution of Oceanography, La Jolla, California
DANIEL SUMAN, Rosenstiel School of Marine and Atmospheric Science, University of Miami, Florida

Ocean Studies Board Staff
SUSAN ROBERTS, Director
JENNIFER MERRILL, Senior Program Officer
DAN WALKER, Senior Program Director
JOANNE BINTZ, Program Officer
ALAN B. SIELEN, Visiting Scholar
ANDREAS SOHRE, Financial Associate
SHIREL SMITH, Administrative Associate
JODI BACHIM, Senior Project Assistant
NANCY CAPUTO, Senior Project Assistant
SARAH CAPOTE, Project Assistant

Appendix D

Water Resources Development Act of 2000
Public Law No. 106-541, of the 106th Congress

SEC. 216. NATIONAL ACADEMY OF SCIENCES STUDY.

(a) DEFINITIONS—In this section, the following definitions apply:

(1) ACADEMY—The term "Academy" means the National Academy of Sciences.

(2) METHOD—The term "method" means a method, model, assumption, or other pertinent planning tool used in conducting an economic or environmental analysis of a water resources project, including the formulation of a feasibility report.

(3) FEASIBILITY REPORT—The term "feasibility report" means each feasibility report, and each associated environmental impact statement and mitigation plan, prepared by the Corps of Engineers for a water resources project.

(4) WATER RESOURCES PROJECT—The term "water resources project"' means a project for navigation, a project for flood control, a project for hurricane and storm damage reduction, a project for emergency streambank and shore protection, a project for ecosystem

restoration and protection, and a water resources project of any other type carried out by the Corps of Engineers.
(b) INDEPENDENT PEER REVIEW OF PROJECTS—

(1) IN GENERAL—Not later than 90 days after the date of enactment of this Act, the Secretary shall contract with the Academy to study, and make recommendations relating to, the independent peer review of feasibility reports.

(2) STUDY ELEMENTS—In carrying out a contract under paragraph (1), the Academy shall study the practicality and efficacy of the independent peer review of the feasibility reports, including—

(A) the cost, time requirements, and other considerations relating to the implementation of independent peer review; and

(B) objective criteria that may be used to determine the most effective application of independent peer review to feasibility reports for each type of water resources project.

(3) ACADEMY REPORT—Not later than 1 year after the date of a contract under paragraph (1), the Academy shall submit to the Secretary, the Committee on Transportation and Infrastructure of the House of Representatives, and the Committee on Environment and Public Works of the Senate a report that includes-

(A) the results of the study conducted under paragraphs (1) and (2); and

(B) in light of the results of the study, specific recommendations, if any, on a program for implementing independent peer review of feasibility reports.

(4) AUTHORIZATION OF APPROPRIATIONS—There is authorized to be appropriated to carry out this subsection $1,000,000, to remain available until expended.

(c) INDEPENDENT PEER REVIEW OF METHODS FOR PROJECT ANALYSIS—

(1) IN GENERAL—Not later than 90 days after the date of enactment of this Act, the Secretary shall contract with the Academy to conduct a study that includes—

(A) a review of state-of-the-art methods;

(B) a review of the methods currently used by the Secretary;

(C) a review of a sample of instances in which the Secretary has applied the methods identified under subparagraph (B) in the analysis of each type of water resources project; and

(D) a comparative evaluation of the basis and validity of state-of-the-art methods identified under subparagraph (A) and the methods identified under subparagraphs (B) and (C).

(2) ACADEMY REPORT—Not later than 1 year after the date of a contract under paragraph (1), the Academy shall transmit to the Secretary, the Committee on Transportation and Infrastructure of the House of Representatives, and the Committee on Environment and Public Works of the Senate a report that includes—

(A) the results of the study conducted under paragraph (1); and

(B) in light of the results of the study, specific recommendations for modifying any of the methods currently used by the Secretary for conducting economic and environmental analyses of water resources projects.

(3) AUTHORIZATION OF APPROPRIATIONS—There is authorized to be appropriated to carry out this subsection $2,000,000. Such sums shall remain available until expended.